AF531558

OUTLINE OF BIOCHEMISTRY

OUTLINE OF BIOCHEMISTRY

Ulag Mahadevan

ANMOL PUBLICATIONS PVT. LTD.
NEW DELHI-110 002 (INDIA)

ANMOL PUBLICATIONS PVT. LTD.
Regd. Office: 4360/4, Ansari Road, Daryaganj,
New Delhi-110002 (India)
Tel.: 23278000, 23261597, 23286875, 23255577
Fax: 91-11-23280289
Email: anmolpub@gmail.com
Visit us at: www.anmolpublications.com

Branch Office: No. 1015, Ist Main Road, BSK IIIrd Stage
IIIrd Phase, IIIrd Block, Bangalore-560 085 (India)
Tel.: 080-41723429 • Fax: 080-26723604
Email: anmolpublicationsbangalore@gmail.com

Outline of Biochemistry

First Edition, 2010

ISBN 978-81-261-4484-6

PRINTED IN INDIA

Printed at Mehra Offset Press, Delhi.

Contents

Preface

Outline of Biochemistry is a resourceful and comprehensive textbook written for graduate and postgraduate students of Bio Science and Engineering of any University. Written in a student-friendly style, the book provides an in-depth knowledge of Biology with involvement of Chemistry. Striking a balance between theory and practice, it covers topics related to pH, water and its solvents, buffers, diseases disorders and metabolism.

This book has been written with the sole objective of providing exposure on various facts of biocompounds and their approach in managing an organization to students. This will enable students in identifying various functional areas where biology and chemistry plays a vital role. The book provides a comprehensive step by step information that is almost essential for better understanding of the concepts covered in the entire book.

Author

Chapter 1

Introduction to pH

OVERVIEW

IONIZATION OF WATER

Sometimes the hydrogen of one water molecule will "jump" to another water molecule:

$$H_2O + H_2O = H_3O^+ + OH^-$$

This proton hopping is called the ionization of water (an ion is a positively or negatively charged atom or molecule). This ionization creates a H_3O^+ and a OH^- molecule. The H_3O^+ is often written as simply H^+. This is because a H_3O^+ is just a H^+ that jumps from one water molecule to another:

$$H_2O = H^+ + OH^-$$

So remember, $H_3O^+ = H^+$

Looking at either of the two chemical equations above, it is important to note that the reverse reaction is also occurring.

How much H^+ and OH^- exist in water? Very, very little! The ratio of either H^+ or OH^- to H_2O in neutral water is 1:1,000,000,000! Since this is such a small amount of either H^+ or OH^-, they rarely meet and neutralize each other.

The equilibrium constant, K_{eq} describes the ionization equilibrium of water:

$$K_{eq} = [H^+][OH^-]$$

Because of this relationship it is important to note that if the $[H^+]$ goes up then the $[OH^-]$ must go down, and vice-versa, for the value for the K_{eq} of water must remain constant. For neutral water, the K_{eq} is 1×10^{-14} M and the concentrations of $[H^+]$ and $[OH^-]$ are each 1×10^{-7} M.

Let's look at that last number without the exponent:

0.0000001 M

This is obviously a very small number. A more manageable way to discuss small numbers such as this is to take the negative logarithm. For the concentration of $[H^+]$, this is called the pH. In this case:

$$-\log(0.0000001\ M) = 7$$

The pH of a solution is simply the negative logarithm of $[H^+]$. The pH of a solution describes the acidity of a solution. Acidic solutions are those with a pH of less than 7 and basic solutions have a pH greater than 7. A solution, like H_2O, with a pH = 7 is neutral. Similarly, the pH could be used to describe a solution in terms of its OH^- concentration. pOH is the negative logarithm of the OH^- concentration.

One useful thing to remember is:

$$pH + pOH = 14.$$

In the body, the pH of blood is 7.4. This corresponds to a $[H^+]$ of about 40 nM. This value can only vary from 37 nM to 43 nM without serious metabolic consequences. pKa In living systems, much of the chemistry involves interactions between acids and bases. Acids are H^+ donors andbases are H^+ acceptors.

$$HA{=}H^+ + A^-$$

Acid and base reactions are made up of conjugate acid-base pairs. Strong acids are those that readily give up a H^+. Strong bases readily accept a H^+. Weak acids do not readily give up a H^+, but will under the right conditions. A weak base is one that does not easily take up a H^+.

The conjugate base for a strong acid is a weak base. In contrast, a strong base has a weak acid as its conjugate. Think of a strong acid as a person with an ugly wig. This person can't wait to get rid of this! So now we have a wig (H^+) and a person without the wig (the weak base). You can understand that the person without the wig doesn't really want to take it back!

However, what about strong base-weak acid conjugate pairs?

$$A^- + H^+ \rightleftharpoons HA$$

Think of it in similar terms: a person in dire need of a wig will grab just about anything. This is our strong base. Now on the right side of the equation we have someone with a wig who really doesn't want to give it up, just as a weak acid does not want to part with its H^+. We can think of a strong acid or base as having a "flip side", which is its weak base or weak acid conjugate, respectively. *How readily an acid gives up its H^+ is expressed by the acid dissociation constant, or K_a:*

$$K_a = \frac{[H+][A-]}{[]}$$

The pK_a is the negative logarithm of the K_a. Strong acids have small pK_as. Looking at the above equation, it can be seen that when $[A^-] = [HA]$, then $K_a = [H^+]$. Then the pK_a = pH. This is the basis for the Henderson-Hasselbalch equation:

$$pH = pK_a + \log\frac{[H+][A-]}{[HA]}$$

This is Henderson-Hasselbalch Equation.

TITRATION CURVES

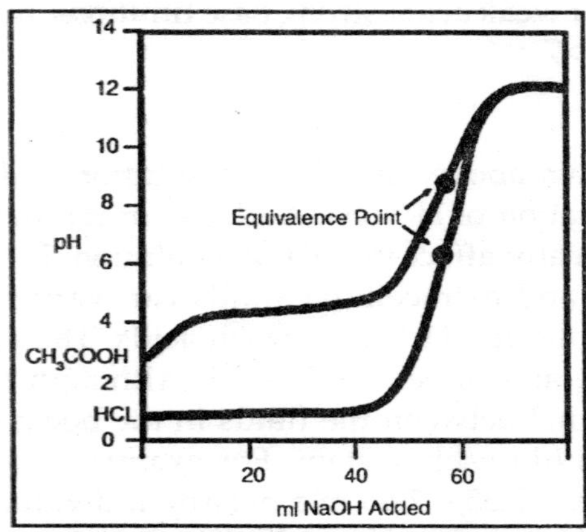

Fig. Titration Curves of Both a Strong and Weak.

Acid with a Strong base: When acids or base are added to a solution, the pH changes. The controlled addition of an acid or base to a solution is called a titration. A titration curve can be made by plotting the pH changes against the volume of acid or base added to a solution. Many titration curves are made by plotting pH on the y axis and the volume of base added to the solution on the x axis. For example, pH changes to acetic acid, CH_3COOH, and hydrochloric acid, HCl, with 0.1M sodium hydroxide, NaOH, is shown to the left. Notice the shape of the curves. The CH_3COOH curve starts out at a higher pH than the HCl curve. This is because CH_3COOH is a weaker acid than HCl. As NaOH is added to the solution of CH_3COOH, the pH rises. However, the addition of NaOH to the HCl solution does not significantly change the pH. It takes less NaOH to change the pH of a weak acid solution than it does to change the pH of a strong acid solution. Eventually all of the acid in both solutions will have reacted with base, such that adding more base quickly increases the pH. This is called the equivalence point.

This is the point where all the acid has reacted with the base, such that further additions of base quickly raise the pH. The equivalence point for a strong acid-strong base titration is pH = 7. For weak acids-strong base titrations, the equivalence point is pH >7.

BUFFERS

As shown above, the pH of a solution is dependent on the concentration of H^+ ions. Addition or removal of H^+ ions, then, can greatly affect the pH of a solution. In the body, the pH of cells and extracellular fluids can vary from pH 8 in pancreatic fluid to pH 1 in stomach acids. The average pH of blood is 7.4, and of cells is 7 – 7.3. Although there is great variation in pH between the fluids in the body, there is little variation within each system. For example, blood pH only varies between 7.35 – 7.45 in a healthy individual.

Large changes in pH can be life threatening. How does the body maintain a constant blood pH? The body uses a buffer system to withstand changes in pH. Buffers are made up of a mixture of a weak acid with its conjugate base or a weak base with its conjugate acid. Remember that an acid donates a H^+. A weak acid does not donate its H^+ as easily. Similarly, a weak base will not accept a H^+ as well as a strong base.

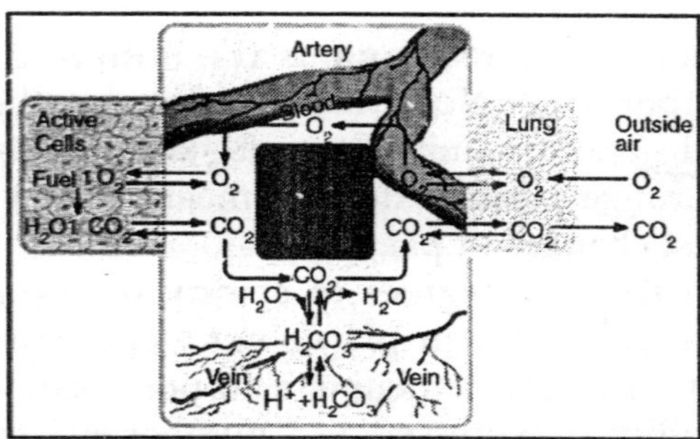

Fig. The Bicarbonate Blood Buffering System

Buffers maintain pH by binding H^+ or OH^- ions. This stabilizes changes in pH. The bicarbonate buffer system maintains blood pH near pH 7.4. The carbonic acid, H_2CO_3, in the blood is in equilibrium with the carbon dioxide (CO_2), in the air. Buffers are most effective in a pH range near its pK_a. This is where the titration curve is most shallow, and where the pH is least affected by added acid or base. For example, in the titration curve of phosphoric acid (another blood buffering system):

$$H_3PO_4 \overset{pK_a1}{\rightleftharpoons} H_2PO_4^- + H^+ \overset{pK_a2}{\rightleftharpoons} HPO_4^{-2} + H^+ \overset{pK_a3}{\rightleftharpoons} PO_4^{-3} + H^+$$

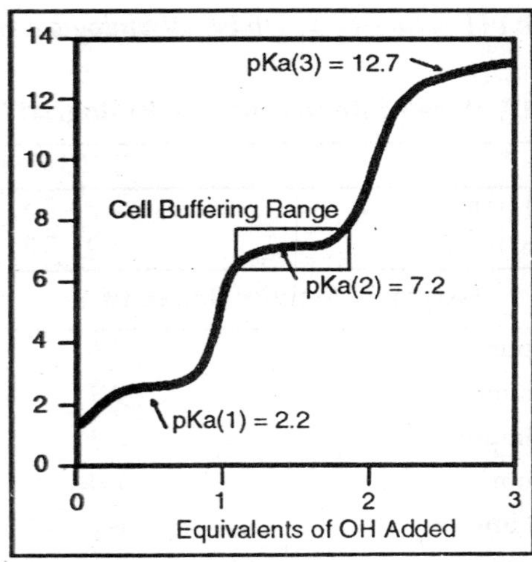

Fig. The Titration of a Polyprotic Weak Acid

The region in the rectangle is the buffering region of biological importance. Note that the slope of the curve is shallow within one pH unit of the pK_a value. This is where the buffering range is most effective, meaning that the buffer is able to resist changes in pH.

In chemistry, the degrees of acidity or alkalinity of a substance are expressed in pH values. The neutral point, where a solution would be neither acid nor alkaline is pH 7. Increasing acidity is expressed as a number less than 7, and increasing alkalinity is a number great than 7. Maximum acidity is pH 0 and maximum alkalinity is pH 14. Because each unit on the scale represents a logarithm, there is 10-fold difference between each unit (pH 5 is 10 times as acid as pH 6 and pH 4 is 100 times as acid as pH 6. Expressed mathematically, pH is logarithm of the hydrogen ion concentration divided into one - Log 1/H+

An acid is a substance that releases hydrogen into a solution. An alkaline substance is one that removes hydrogen from a solution. Most body functions occur at only certain levels of acidity or alkalinity, as well as, many enzymes and chemical reactions work best at a specific pH. A minute change in pH can have a profound effect on body function.*

Some of the pH values of a number of biological materials are given below:

Table A - pH Values of Representative Biological Materials

Material	pH value
Blood, normal limits	7.3 - 7.5
Blood, extreme limits	7.0 - 7.8
Enzymes, Activity Range of	
Amylopsin, optimum	7.0
Erepsin, optimum	7.8
Invertase, optimum	5.5
Lipase, optimum	7.0 - 8.0
Maltase, optimum	6.1 - 6.8
Pepsin, optimum	1.5 - 2.4
Trypsin, optimum	8.0 - 9.0

Fruit Juices	
Apple	3.8
Banana	4.6
Grapefruit	3.0 - 3.3
Orange	3.1 -4.1
Tomato	4.2
Gastric juice (adult)	0.9 - 1.6
Milk (cows)	6.2 - 7.3
Plants (Extracted Juice)	
Alfalfa tops	5.9
Carrot	5.2
Cucumber	5.2
Peas, field	6.8
Potato	6.1
Rhubarb, stalks	3.4
String beans	5.2
Sweat	4.5 - 7.1
Saliva	6.2 - 7.6
Urine (human)	4.2 - 8.0
Tears	7.2

PH - POTENTIAL OF HYDROGEN DESCRIPTION

What Is Hydrogen

Hydrogen (H) is the lightest and simplest element. A Hydrogen atom is made up of one proton and one electron. At room temperature, Hydrogen is gas - as it cools it becomes liquid. Hydrogen is one of 9 essential macronutrients and is a major component of organic molecules. Hydrogen supplies building blocks of all organisms whether plant, animal or human. Each water molecule contains two Hydrogen molecules and one Oxygen molecules (H20).

Ideal pH Values

When pH = 8: There are 6 billion H+ ions in solution.
When pH = 7: There are 60 billion H+ ions in solution.
When pH = 6: There is 600 billion H+ ions in solution.

Ideal pH Values Of Blood, Urine and Saliva

- Arterial Blood – 7.40 – 7.45
- Capillary – 7.35 – 7.45
- Venous – 7.30 – 7.35
- Saliva – 6.5
- Urine – 6.8

pH Values Of Body

Saliva – (6.0 – 7.4)	Skeletal Muscle - (6.9 - 7.2)
Gastric Secretion - (1.0 - 3.5)	Heart - (7.0 - 7.4)
Pancreatic Secretion - (8.0 - 8.3)	Liver - (7.2)
Bile - (7.8)	Brain (7.1)
Small Intestine Secretion - (7.5 - 8.0)	
Urine - (4.5 - 8.0)	Arterial Blood - (7.4 - 7.45)
Capillary Blood - (7.35 - 7.4)	Venous Blood - (7.3 - 7.35)
Feces - (4.6-8.4)	

CLINICAL ACIDOSIS AND ALKALOSIS

The blood pH is supposed to remain "practically constant" according to our best text books. The physician who starts to check his patients, however, cannot occur. Instead of the normal figure of pH 7.3 to 7.5 he finds it varying from 7.2 to 7.8 (1).

The measurement of pH is attended with difficulty because the blood loses carbon dioxide (CO_2) and thereby rises the pH rapidly after it has been withdrawn. In two minutes a rise of.2 may occur (1). pH metres are available where the electrodes are built into a syringe so the pH may be read directly before the blood is exposed to air. (National Technical Labouratories, Pasadena, California).

The importance of clinical pH testing was illustrated by the following report of a pharmacist. He had been supplying for a number of months a capsule formula of aspirin, caffeine, phenacetin and ammonium chloride to an arthritic patent where the remedy had been successful in controlling arthritic pains. The capsule dispensed was a standard formulas of a pharmaceutical house, with a specific catalog number. The patient returned two days after receiving a renewed supply of capsules, complaining that they were ineffective, and different from the previously used kind. On checking the records, it was found that the original formulas had been revised, and the supposedly unnecessary ammonium chloride had been omitted. On supplying capsules of ammonium chloride alone, the patient reported the usual successful results.

What was the biochemistry behind this: Simply that a high pH tends to throw calcium out of solution in the body fluids, and bursitis, arthritis, neuritis, lumbago, sciatica, and a host of other painful syndromes develop as a consequence. Cheilosis (lips redden, fissures at the angles) may occur and herpes simplex, and other virus types of diseases become active. The low calcium bicarbonate in the body fluids seems responsible, Vitamin F and Vitamin D as well as the Vitamin C complex are also factors, their deficiency aggravates the situation. Allergies become acute, the calcium deficiency aspect of allergic sensitivity is well known. Normal rates of wound healing are greatly reduced, ulcers then too become static.*

Ammonium chloride is a natural constituent of gastric juice, calcium chloride is the form of calcium found in arrowroot starch, so both of these materials are physiological therapeutic agents, factors common to body biochemistry and not new and foreign substances to the physiological economy (as are synthetics and antibiotics).

It is described that Ammonium Chloride is useful in the treatment of bronchial affections, hepatic congestion, pelvic cellulitis, muscular rheumatism, gout, sciatica, chronic glandular enlargement, hemicrania, senile gangrene, dysmenorrhea, leucorrhea.

Calcium Chloride is listed as useful in the treatment of hemorrhage, hemorrhagic endometritis, menorrhagia, erytherma nodosum, tetanus, spasmophillia, blackwater fever, hay fever, asthma, hemophilia, albuminuria, nephritis, typhoid, coryza, tuberculosis, osteomalacia, scrofula, rachitis, arthritis, spasm of glottis, infant convulsions, urticaria, pruritis, among others.

All of these effects are no doubt in the main simply accomplished by correcting the unbalanced state of the buffers and mineral salts in the tissue fluids. It will be noted that Vitamin C complex is a desirable synergist in most of the conditions listed. Gout is mentioned as one situation where ammonium chloride may help. Recently it has been discovered that the juice of cherries contains some active agent effective in gout, 4 to 6 ounces of the juice (preferably unsweetened, but may be canned or cooked form) promoting relief with in a few days with consistent improvement.

This is the first time that the possibility of gout being a deficiency disease has been suspected. We can say that the ingestion of sodium compounds particularly sodium phosphate - is definitely aggravative. Gout patients should see that their diet contains a preponderance of potassium. Sodium promotes the precipitation of gouty concretions of sodium urate in the tissues.

A very important aspect of alkalosis is the fact the habitual use of milk of magnesia promotes the condition. It also tends to cause epistaxis (nosebleed - can be secondary to nasal and sinus issues; scarlet fever or typhoid; arteriosclerosis; hypertension and anemias), very specifically. You will be surprised how many of the patients who ask for a remedy for nosebleed are using milk of magnesia as a laxative.

Citrus fruits being of a highly alkaline ash, with a high content of organic acid (citric), need special attention. The first and immediate effect of ingestion of a few ounces of grapefruit or lemon juice is to lower (acidity) the blood pH. The patient with alkalosis feels temporarily better. Later in the day, after the citric acid has been destroyed by oxidation as a fuel (it is classed as carbohydrate) the aggravation of the alkaline state become apparent. The temporary effect of the citric acid is to cause calcium to be picked up - no doubt from some bone reserves - and after the pH change this calcium is deposited elsewhere - in physiologically undesirable spots. The treatment for bursitis obviously becomes simple. No longer can we recommend X ray treatment with equanimity. Let the physiological treatment be the preferred method.

Guanidine, a fatigue and tissue poison, and end product of the breakdown of creatine (combated by Vitamin E complex which stops the loss of creatine from the tissues), is the most potent organic alkaline substance know. It specifically precipitates calcium (recall our comment that it is diffusion from tired heart muscle into coronary vessels is the cause of precipitation therein of calcium), is normally reconverted into creatine by the influence of the parathyroid hormones, and with the assistance of the thyroid (2). Blood guanidine levels rise eight fold after parathyroidectomy (3).*

We can see where the guanidine effect as a fatigue poison is contributory to the arthritis, sciatica, etc., the patient often tell us how fatigue aggravates his state of misfortune. Local inflammation also can release guanidine, its presence insure a spastic state of blood vessels, contributing to gangrene, of necessity.

There seems no doubt that the effect of oxygen metabolizing vitamins - the Vitamin C and Vitamin E complexes - is vital in combating these morbid reactions by preventing the degradation of tissue elements into guanidine. Where Vitamin E deficiency has caused the tissue demand for oxygen to rise up to 250%, of normal (4), the augmented release of end products certainly includes guanidine, and we see here how the find of Dr. Shute of London, Ontario, that Vitamin E reduces capillary hemorrhage may be rationalized.*

A vicious cycle is set up, the more oxygen demand the less supplied by reason of the constrictive effect of guanidine on blood vessels. Degeneration, rupture of capillaries is inevitable. As Vitamin E promotes more oxygen supply, that is why it also is good in gangrene (especially the diabetic type) and in "capillary fragility". (The true antifragility vitamin is Vitamin P, which provides a special calcium to promote collagen formation.)

Vitamin F opposes the toxic effect of guanidine by providing more of the diffusible calcium (from the colloidal blood reserves) that is precipitated out of the body fluids by guanidine, thereby ameliorating the spastic, nervous and irritative (allergic) reactions.

The thyroid hormone physiologically promotes the resorption and dissolution of protein structures that have reached the end of their physiological cycle (the theory of the dynamic state of living tissue, replaced and rebuilt at specific time intervals). That is why in children, thyroid deficiency is the cause of delayed development. In the adult, a hyperactive thyroid may again cause toxicosis by promoting more tissue poisons than the eliminative system can tolerate. In old people, this may be a critical situation, thyroid sometimes becoming a violent poison.

Only because the normal synergists are not available to maintain the desirable complete cycle of activity. We can see that Vitamin F is vital to the prevention of this toxicosis. It tells why the thyroid puts its secretion into the blood after ingestion of Vitamin F to the extent of a doubling of the iodine content of the blood (5). The normal activity of the gland had been blocked by reason of Vitamin F deficiency. In this blocking, we have apparently the explanation for prostate hypertrophy. Once the thyroid is permitted to secrete, its hormone makes quick work of the fibrous tissue collected in the prostate. This also explains why we consider the Vitamin F complex, and Vitamin F2 the most friendly vitamins for the older person. The Vitamin F2 is a more highly developed compound of the basic F (a sensitized fatty acid), in which this fatty acid is combined into a phospholipid molecule that appears to have a specific function of catalyzing protective (Insulating) layers in nerve tissue and cell nuclear structure. Thereby protecting the basic controls of metabolic activity. Loss of appetite and wasting disease suggests the possible need for this vitamin. Appetite restoration is immediate where the deficiency exists (In adults and children alike.)

There seems to be as much clinical acidosis as alkalosis (1). The effects of acidosis may be aggravated by a diabetic state where organic acids cannot be oxidized normally, instead must be combined with reserve alkaline salts and excreted (5).

Where sodium bicarbonate is administered to relieve diabetic acidosis, the urine should be watched and dosage stopped when a neutral is reached, according to this authority.

Normally, blood glutamine provides ammonia for this purpose. Beet root and leaf are best vegetable sources of glutamine, is very high if grown on high ammonia content soils. (Plant cells form glutamine to dispose of excess ammonia).

When abnormal blood pH is found, it must be considered only as a sign that something is wrong. The symptoms may be entirely different with the same abnormal pH figure in different patients, in view of the fact that there is an infinite number of combination of chemical situations that could alter pH - as many as there are different acids and alkalies. Just as a plump patient may be bloated with gas, water logged, fat, or hyper muscular. It is a preliminary classification, not a final.

The pH of the blood is a resultant of the forces acting upon it. The regulation of the blood CO_2 (carbon dioxide) by respiration control Centre of the brain is a major factor. The parathyroid by eliminating the alkaline guanidine, and the kidney, eliminating mineral salt (both acid and alkaline as may be required) no doubt exercises a supplementary control.

The clinical value of vinegar is more or less well known as an alternative of merit. Creatine is methylguanidine acetic acid, and is often craved by the patient, very logically it would seem.

Cider vinegar being the preferred form, no doubt the malic acid of the apple is a factor, calcium malate being a catalyzer of polyphenol detoxification.

ACIDOSIS AND ALKALOSIS INFORMATION

Regulatory Factors

The following factors tend to promote normalization of pH through physiological mechanisms. They do not supply mineral elements which directly influence body chemistry. To directly influence pH values through acid or alkaline mineral administration, see: Alkalosis or Acidosis.

Normalization of the function of the glands and kidneys which regulates pH balance should be the primary consideration. However, in most cases specific support of either acid or alkaline mineral therapy is necessary where immediate results are to be obtained, particularly in acute situations.

Primary

- *Renatrophin PMG*: Kidneys are thought to be the most important organ in acid-base balance regulation.
- *Drenatrophin PMG*: Adrenal support regulation of sodium-potassium-chloride balance.
- *Thytrophin PMG*: Thyroid supports regulation of calcium-magnesium-phosphorus balance.

Normally the endocrine glands, along with the kidneys regulate the pH of the blood, just as the pancreas with it insulin regulated the blood sugar levels. The sex glands, along with the adrenals, seem to be the main endocrines involved. Particularly after menopause or male climacteric there may be serious changes in the body's economy whereby acid-base disorders develop, which may be aggravated, if not caused by dietary circumstance - just as diabetes is aggravated by a high sugar intake.*

Synergistic Supplements

- *Pituitrophin PMG: Pituitary is the hormonal control of* endocrine system
- *Ovatrophin PMG*: Specific cell activator
- *Utrophin PMG*: Specific cell activator
- *Prostate PMG*: Specific cell activator
- *Orchic PMG*: Specific cell activator

ALKALOSIS

PHYSIOLOGICAL CONSIDERATIONS

The calcium may precipitate out of body fluids (with excess tissue calcium) where alkalosis exists. Increased contractility of muscles noted. There may be calcium deficiency symptoms, but cannot assimilate calcium in any form.

The neuritic pains which may develop are no doubt due to calcium crystals (calcium carbonate) that form at nerve endings which accumulate and disperse according to the variations of the pH of the blood that follow dietary levels of acid and alkaline food ingestion, unable to be compensated for by a depleted endocrine system.

FREQUENTLY OBSERVED ENTITIES

Menopausal Disorders; Allergic States; Osteoarthritis; Renal Calculi; Prostatitis; Bursitis; Bronchitis; Asthma; Chronic indigestion; Coronary Sclerosis; Hypercholesterolemia.

COMMON CLINICAL SITUATIONS

- Heavy slow pulse
- Stiffness of joints
- Symptoms after resting
- Skin Symptoms)
- Blood sludge "thickening"
- Feel worse after eating

PREDISPOSING FACTORS

Endocrine insufficiency (kidneys, thyroid, adrenals, gonads, etc):

- Excess carbohydrate intake (citrus fruits, sugar, etc.)
- Excessive alkaline ash foods
- Mental Stress
- Fluid loss (perspiration, diarrhea, vomiting, etc)
- The following supplements tend to combat the effects of alkalosis

Primary

- *Cal-Amo*: Chloride ion source.
- *Phosfood* Liquid: Orthophosphoric acid source.
- *Soy Bean Lecithin*: Phosphorus and choline source.
- *Betaine Hydrochloride*: Hydrochloric acid source.

Secondary

- *Cataplex V G*: Acetic acid metabolism
- *Bio-Dent*: Bone-source of phosphorus
- *Calcifood*: Bone source of phosphorus
- *Biost*: Phosphatase source (enzyme)
- *Prost-X*: Phosphatase source (enzyme)

METABOLIC ALKALOSIS

The pH scale runs from 0-14 with 7.0 be a neutral pH and with a pH of 6.4 to 6.8 being the most suggested when considering your biological terrain first thing in the morning (urine and saliva) - it will change to alkaline numbers throughout the day with 7.0 - 7.8 being the norm a couple hours after eating.

Causes Of Metabolic Alkalosis

All diuretic therapies except for the carbonic anhydrase inhibitors. These diuretics cause a tremendous increase in the concentration of Na+ ions brought into the distal and collecting tubules. This lead to rapid reabsorption of these sodium ions and creates an enhance hydrogen ion secretion due to the Na+-H+ exchange mechanism. Na+ get reabsorbed and H+ get suddenly excreted.

Excessive ingestion of alkaline drugs (Rx of gastritis and peptic ulcers)

WATER, PH, AND NON-COVALENT BONDING

WATER

Water is essential for life. It covers 2/3 of the earth's surface and every living thing is dependent upon it. The human body is comprised of over 70% water, and it is a major component of many bodily fluids including blood, urine, and saliva. What accounts for the ubiquitous use of water in living systems? If we take a step back and consider the structure of water and compare it to another substance, methane, we can understand the unique properties water possesses that make it well suited for biological systems.

Water (H_2O) is made up of 2 hydrogen atoms and one oxygen atom, with a total atomic weight of 18 daltons. The structure of the electrons surrounding water is tetrahedral, resembling a pyramid. For comparison, Methane (CH_4) is made up of one carbon and 4 hydrogens. Note that methane is similar to water in that it weighs 16 daltons and also has a tetrahedral structure, yet has very different physical properties.

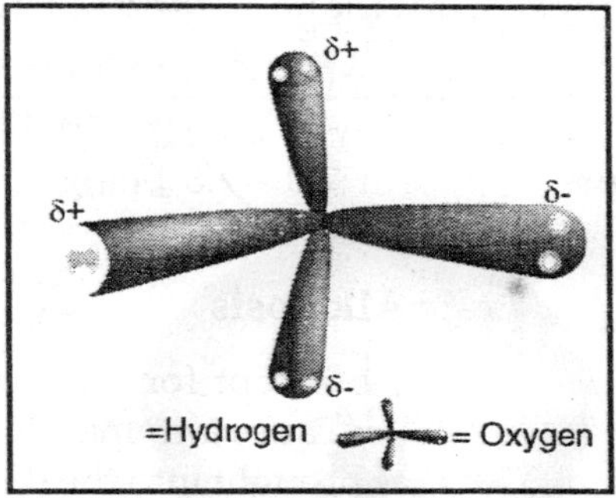

Fig. Molecular Geometry of Water: The Orbitals are Pyramidal, Yet the Atoms form a Bent Shape.

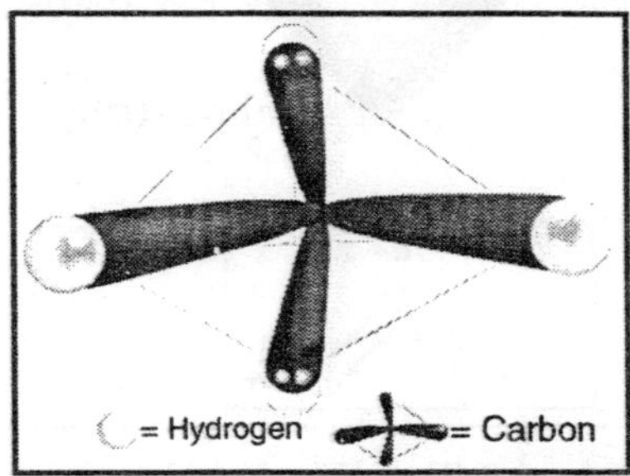

Fig. Molecular Geometry of Methane: Pyramidal

All the vertices on the methane that form the points of a pyramid are occupied by a hydrogen. However, only two vertices in water are occupied with hydrogens. The other two vertices are each occupied by a lone pair of electrons. This fact causes the water molecule to have a bent molecular shape. Oxygen is highly electrophilic (electron loving).

This means that even though the oxygen in water is bound to each of the hydrogens by a covalent bond (sharing a pair of electrons), the oxygen "pulls" the shared electrons closer to itself. This unequal sharing of the electrons in the O-H bond in water causes the hydrogens to have a partial positive charge (positive dipole), and the oxygen has a partial negative charge (negative dipole). Water is called a polar molecule because it has a positive side and a negative side, called a dipole moment. In contrast, the carbon in methane shares electrons equally with the 4 hydrogens. Methane's does not have a dipole moment, and in contrast with water is non-polar.

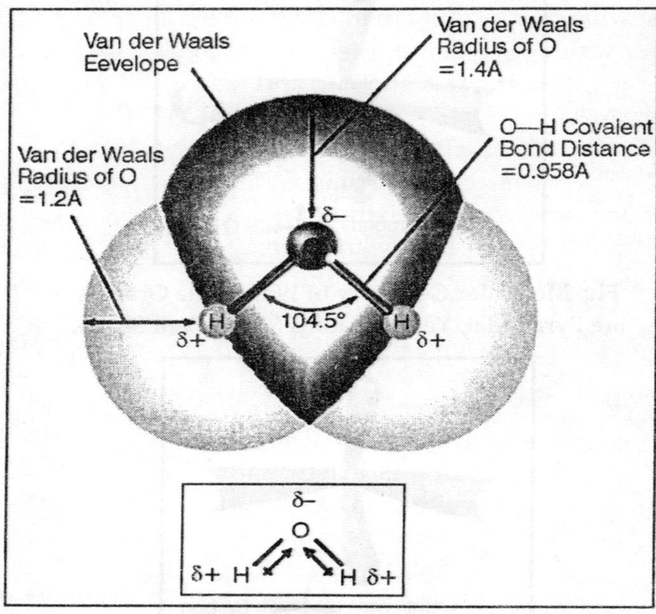

Fig. The "Magnet-like" Dipole Moment in Water

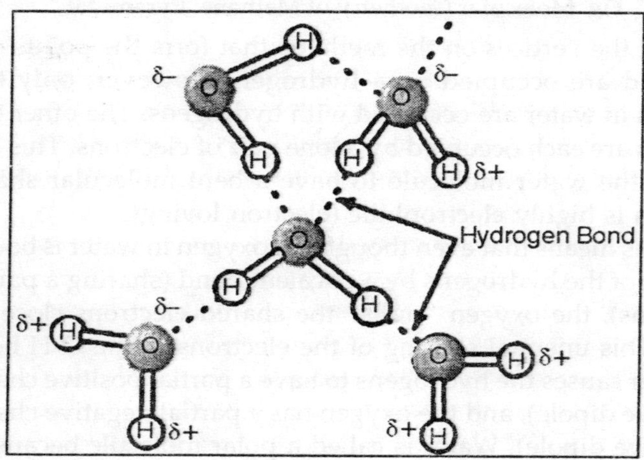

Water molecules, then, have both a partial positive and a partial negative charge. Since opposites attract, this means that water molecules are attracted to each other, like the positive and negative ends of a magnet. Because of these attractive forces, water molecules are in close proximity to one another, making water very dense. Methane molecules do not have a dipole attraction for one another, and therefore and are spaced farther apart.

Therefore, despite its similar size and mass to water, methane is much less dense than water. This is the reason that under room temperature situations, water exists as a liquid while methane is a gas. We know that water can be converted to its gaseous phase, steam, but only by applying a lot of energy in the form of heat to disrupt the large attraction of the water molecules for one another.

Hydrogen Bonding

Water molecules are bound together through hydrogen bonds. Hydrogen bonds arise when a hydrogen that is covalently bound to either an oxygen, sulfur, or nitrogen atom nears an electrophilic atom, such as oxygen. The electrophilic atom "pulls" the hydrogen closer to itself. The end result is that the hydrogen is now shared (unequally) between the atom to which it is covalently bound and the electrophilic atom to which it is attracted (O-H...O).

Since water is made up solely of hydrogens covalently bound to oxygens and electrophilic oxygens, it is easy to see why there are so many hydrogen bonds in water. The molecules in water are in constant motion and the hydrogen bonds are constantly being broken and reformed. In fact, the average hydrogen bond lasts only a one-trillionth of a second (10^{-12} s)! Each hydrogen bond has an average energy of 20 kJ/mol. This is much less than an O-H covalent bond, which is 460 kJ/mol. Even though an individual hydrogen bond is relatively weak, the large number of hydrogen bonds that exist in water which pull the molecules together give water its special density and phase transition properties. This is much like the story of Gulliver's Travels, when Gulliver awoke on the island of Lilliput to find himself tied down with many tiny strings. Each string was individually weak, but together they were enough to bind Gulliver.

On average, a liquid water molecule will have 3 out of 4 possible hydrogen bonds. This is because the molecules in liquid are in constant motion. In ice, water will form all 4 possible hydrogen bonds because the molecules in a solid are essentially locked in place. Water molecules are further apart from one another in ice than they are in the liquid state. Ice is therefore not as dense as liquid water.

This is why icebergs float in the ocean and why ice floats in a beverage. This contrasts greatly from the normal relationship between a compound's solid and liquid forms, with the solid form usually being of a higher density. While the unusual phase density properties of water might not seem too important in biological reactions, it is very important for life on earth. If ice did not float, then in very cold bodies of water ice would sink, removing its insulating effects from the water's surface. Lakes, rivers, and major portions of the oceans thus would freeze completely from the bottom up. The world would be a much different place, and much less suited for life.

Water is a Good Solvent

The polar nature of water, with its partial positive and partial negative dipole, allows it to dissolve charged molecules (ions) easily. Water is thus an excellent solvent for charged compounds. The positive side of water surrounds negatively charged molecules, and the negatively charged side of water surrounds positively charged molecules.

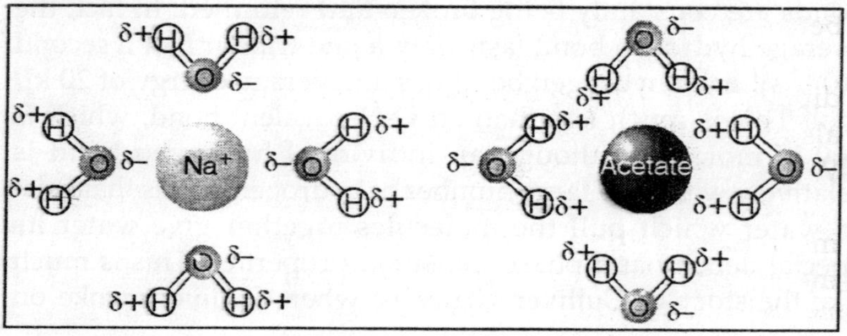

In this way, water makes "solvation shells" around ions. Water can also readily dissolve other polar molecules, even if they are not positively or negatively charged. The solvent properties of water allow for dissolved metals and buffering systems that are very important for the workhorses of life, enzymes. However, the saying "oil and water don't mix" is true—water cannot dissolve oil. This is because oily substances are non-polar. Non-polar substances (which lack dipoles) are also called *hydrophobic* (water fearing). Hydrophobic substances gather together to exclude water as best they can. This is why you see oil droplets in water. This is also important for the stability and structure of enzymes.

NON-COVALENT BONDING

Non-covalent bonds are not at strong as covalent bonds, but they are important in the stabilization of molecules. In contrast to covalent bonds, non-covalent bonds do not share electrons.

Noncovalent bonds include:

- Electrostatic interactions
- van der Waals forces
- Hydrophobic interactions
- Hydrogen bonds

Electrostatic interactions are formed between positive and negative ions. The bond is non-directional, meaning that the pull of the electrons does not favour one atom over another. An example is NaCl, which is formed between the positively charged Na^+ ion and the negatively charged Cl^- ion. The average strength of an electrostatic interaction is 15 kilojoules/mol (kJ/mol). The bond strength lessens when the distance between the two ions increases.

Van der Waals forces are weak forces between temporary dipoles. These forces may be attractive or repulsive. They are also non-directional. The average bond energy for van der Waals forces is on the order of 10 kJ/mol.

Hydrophobic interactions result when non-polar molecules are in a polar solvent, e.g. H_2O. The non-polar molecules group together to exclude water (hydrophobic means water fearing). By doing so they minimize the surface area in contact with the polar solvent.

A hydrogen bond results when a hydrogen atom that is covalently bound to an electronegative atom (e.g. O, N, S) is shared with another electronegative atom. A hydrogen bond is directional towards the electronegative atom. An example of this is the hydrogen bonds formed in water. Hydrogen bonds are constantly being made and remade. The half-life of a hydrogen bond is about 10 seconds. A hydrogen bond has an energy of 21 kJ/mol.

As stated above, non-covalent bonds are not as strong as covalent bonds, but the additive effect of many non-covalent bonds can stabilize a molecule. Non-covalent bonds are very important in the structure of proteins.

Bond type	Energy (kJ/mol)
Covalent, e.g. C-C	350
Electrostatic	15
van der Waal's	10
Hydrogen	21

TEST AND PROBLEMS[H]

pH Questions

Below are a few sample questions to elucidate the concepts discussed in the book.

- A simple pH problem
- pH of a weak acid
- A buffer problem
- A more involved buffer problem

A Simple pH Problem

What is the pH 0.1 M HCl?

Answer: The pH of a solution is the negative logarithm of the concentration of H^+ ions. In this case, HCl is a very strong acid. HCl completely dissociates to H^+ and Cl^-.

$$HCl \rightarrow H^+ + Cl^-$$

The concentration of HCl is 0.1 M, if all of it dissociates to H^+, then

$$[H^+] = 0.1$$
$$-\log(0.1) = 1.26$$
$$pH = 1.26$$

Remember that a pH of 7 is neutral, and that the pH scale is logarithmic, meaning each pH unit represents a factor of 10. This HCl solution has a difference of 5.74 pH units from water (7 - 1.26). Taking the inverse log of 5.74 means that this HCl solution is more than 500,000 times as acidic as water.

pH of a Weak Acid Solution

What is the pH of a 0.05 M solution of acetic acid? (The K_a of acetic acid = 1.8×10^{-5})

Answer: Determining the pH of a weak acid solution is more complicated than for a strong acid. This is because a weak acid does not completely dissociate.

$$CH_3COOH \rightleftharpoons H^+ + CH_3COO^-$$

To simplify matters, it is helpful to construct a grid:

Reaction	CH_3COOH	H^+	CH_3COO^-
Initial	0.05 M	0 M	0 M
Change	(–x) M	+ x M	+x M
Equilibrium	(0.05–x) M	x M	x M

To determine the $[H^+]$, we must employ the K_a of acetic acid, which is 1.8×10^{-5}.

$$Ka = \frac{[H+][CH_3COO-]}{[CH_3COOH]} = 1.8 \times 10{-}5$$

Substituting the equilibrium values in the table above, we get:

$$1.8 \times 10{-}5 = \frac{(x\ M)(x\ M)}{(0.05 - x)\ M}$$

Since acetic acid is a weak acid, we can assume that very little of it will dissociate. *This simplifies the above equation to:*

$$1.8 \times 10^{-5} = \frac{(x\ M)(x\ M)}{0.05}$$

This can be easily solved:

$9 \times 10^{-7}\ M = x^2\ M$

$x = 9.5 \times 10^{-4}\ M$

$x = 0.00095\ M$

Now that we have the [H+], we can solve for pH:

pH = –log[0.00095]

pH = 3

Buffer Problem

Question 1: What is the ratio of [lactate]/[lactic acid] at pH = 5? (The pK_a of lactic acid = 3.85)

Answer: To solve this problem, we need to utilize the Henderson-Hasselbalch equation:

$$pH = pKa + \log \frac{[H+][A-]}{[HA]}$$

Table in the book lists the pKa of lactic acid as 3.85. Now just substitute into the equation:

$$5 = 3.85 + \log \frac{[lactate]}{[lactic\ acid]}$$

$$1.15 = \log \frac{[lactate]}{[lactic\ acid]}$$

$$14 = \frac{[lactate]}{[lactic\ acid]}$$

There is 14 times as much lactate as lactic acid at pH 5.

Question 2: What happens if 2 ml of 100 mM HCl is added to 100 ml of a 10 mM solution of phosphoric acid at pH = 7?

Answer: This is a more involved problem. The first thing to Realise is that phosphoric acid is polyprotic, meaning it has more than one H+ to donate, and therefore more than one pKa.

$$H_3PO_4 \underset{}{\overset{pK_a1}{\rightleftharpoons}} H_2PO_4^- + H^+ \overset{pK_a2}{\rightleftharpoons} HPO_4^{-2} + H^+ \overset{pK_a3}{\rightleftharpoons} PO_4^{-3} + H^+$$

Since this problem is at pH = 7, it is the pKa2 value that is important, for this is the pKa nearest the pH.

First we need to find the ratio of,

$$[HPO_4^{-2}]/[H_2PO_4^-].$$

Again, use the Henderson-Hasselbalch equation:

$7 = 7.2 + \log[HPO_4^{-2}]/[H_2PO_4^-]$

$0.63 = [HPO_4^{-2}]/[H_2PO^{4-}]$

$0.63[H_2PO^{4-}] = [HPO_4^{-2}]$

We know we have a total of: (0.01 moles/L) × (0.1 L) = 0.001

moles phosphoric acid that at pH = 7 is made up of HPO_4^{-2} and H_2PO_{4-}. (Note 10 mM = 0.01 moles/L and 100 mL = 0.1 L)

$$0.001 \text{ moles} = 1 \text{ mmole} = HPO_4^{-2} + H_2PO^{4-}$$

We know from above that $[HPO_4^{-2}] = 0.63[H_2PO^{4-}]$

so 1 mmole=$0.63[H_2PO_4-] + [H_2PO_4-] = 1.63[H_2PO_4-]$

$$[H_2PO_4-] = 0.6 \text{ mmoles}$$

and so then $[HPO_4^{-2}] = 0.4$ mmoles

So far, this problem is much like the previous one. Now let's answer the question about the addition of 1 ml of 0.1 M HCl. From the high pK_a2 value, we know that phosphoric acid is a weak acid. It will react completely with the HCl (which is the same as H+ in solution):

$$\overset{pK_a3}{HPO_4^{2-} + H^+ \rightleftharpoons H_2PO_4-}$$

By adding 2 ml of 0.1 M HCl, we are adding:

$$(1 \times 10^{-3} \text{ L}) \times (100 \text{ mmoles/L}) = 0.2 \text{ mmoles } H^+$$

If we set up the grid as shown above,

Reaction	HPO4-2	H+	H2PO4-
Initial	0.6 mmoles	0.2 mmoles	0.4 M
Change	–0.2 mmoles	–0.2 mmoles	+ 0.2 M
Change after equilibrium	0.4 mmoles	0 mmoles	0.6 M

What is the pH of the phosphoric acid solution after the addition of HCl? Again, use the Henderson-Hasselbalch equation to solve,

pH = 7.2 + log(0.6/0.4)

pH = 7.37

Note that this is not a big change. This is because of phosphoric acid is a buffer near its pKa and as such is able to resist changes in pH.

Chapter 2

Buffers

DEFINITION

A solution that contains a weak acid and its conjugate base has the special property of being able to resist changing its pH when either a base or an acid is added to it. Such a solution is called a buffer because the solution is protected, or buffered, from pH changes even when H_3O^+ or OH- ions are added to the solution.

Buffers help biochemists study biomolecular reactions in the Labouratory by stabilizing the pH of solutions used for experiments. Fluids inside biological organisms must be maintained at a constant pH for the health of the organism. That is because these fluids, whether blood, cytosol, lymph, bile, etc., contain enzymes that carry out the reactions that support life, and each of these enzymes only has a certain pH range within which it can function effectively. This is called as the pH dependency of the enzyme.

If the pH is too high or too low, the amino acid residues important for catalysis might be in the wrong protonated state for catalysis, or the enzyme's conformation could change, rendering it nonfunctional. Shown at right is a graphic for a fictional enzyme that is very similar to the pH dependence of trypsin. Other enzymes in the same organism might have a very different dependence. For example, pepsin has optimal activity at a pH of 2, because it is a digestive enzyme and has to be active in acidic digestive fluids. If the pH of biological fluids is so important, how does an organism make sure its fluids are all kept at the proper pH for the correct functioning of all of its enzymes, even if the organism ingests material that is acidic or basic? The answer is that all the biological fluids are buffered. Buffers are solutions that are resistant to changes in pH, even if acid or base is added to the solution.

Because enzymes are so pH-sensitive, not only for their activity but also for their stability in solution, buffers are very important for biochemical research. In a lab, researchers keep all of their proteins in buffered solutions. Therefore, beginning biochemists need to understand how buffers work in the body and how to make them in the lab. This difficult concepts section is a primer to help you grasp buffers to a level of understanding needed for introductory biochemistry. *Go through the subsections in order to get the best instruction*: In chemical and biological processes the pH value is an essential parametre.

Slight changes in pH cause negative results between the reaction partners. It is therefore necessary to maintain a correct pH and to protect the system from internal and external chemical influences. This work is performed by buffer substances, which assure a pH in a specific range. Dissociation constants, relationship between pH, pKa and the buffering capability are mainly explained in the Henderson-Hasselbach-Equitation. Buffers have a characteristic buffer capacity, buffering range and dissociation constants at defined temperatures. Temperature shows effects on the pH of the buffer. These differences could be quite small but could still cause significant discrepancies in biological systems. Good's Buffers are buffering substances which were selected and described by Norman Good and his collegues in 1966.

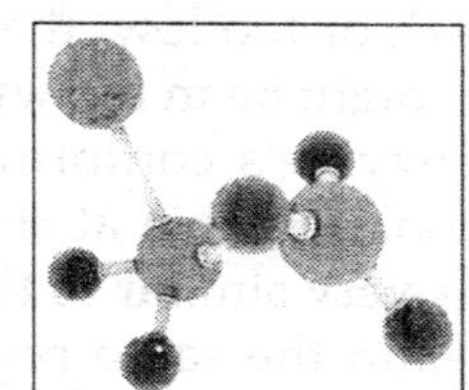

Sinus offers a selected range of buffers, which are widely used for research in biochemistry and biotechnology.

The main data are shown in the following table: A typical certificate of analysis for the common Good buffer Tris -Sinus Code 55065- is also given.

BUFFERS CHARACTERISTICS

Buffer substance	Chemical name	Sinus-Code	M.W.	Buffering range (pH)	pKa (20°C)
		55001	182,2	6,1-7,5	6,90
AMPD					
AMPSO					
BES					7,15
BICINE	N,N-Bis-(2-hydroxyethyl)-glycine	55105	163,2	7,6-9,0	8,35
BISTRIS	Bis-(2-hydroxyethyl)-imino-tris-(hydroxymethyl)-methane	55107	209,2	5,8-7,2	6,4
CHES	2-(N-Cyclohexylamino)-ethane sulfonic acid	55109			
HEPES	4-(2-Hydroxyethyl)-piperazine-1-ethanesulfonic acid	55030	238,3	6,8-8,2	7,55
HEPPS	4-(2-Hydroxyethyl)-piperazin-1-propanesulfonic acid	55110	252,3	7,3-8,7	8,1
HEPPSO	4-(2-Hydroxyethyl)-piperazin-1-(2-hydroxyl)-propanesulfonic acid	55111	268,3	7,1-8,5	7,9
MES	2-Morpholinoethanesulfonic acid	55035	195,2	5,5-6,7	6,15
MOPS	3-Morpholinopropanesulfonic acid	55040	209,3	6,5-7,9	7,2
MOPSO	3-Morpholino-2-hydroxy-propane- sulfonic acid	55042	225,3	6,2-7,6	6,95
PIPES	Piperazine-1,4-bis-(2-ethane)-sulfonic acid	55175	302,4	6,1-7,5	6,80
TAPS	N-[Tris-(hydroxymethyl)-methyl]-3-aminopropane sulfonic acid	55140	243,3	7,7-9,1	8,55
TAPSO	N-[Tris-(hydroxymethyl)-methyl]-3-amino-2-hydroxypropanesulfonic acid	55142	259,3	7,0-8,2	7,7
TES	N-[Tris-(hydroxymethyl)-methyl]-2-amino-ethanesulfonic acid	55180	229,3	6,8-8,2	7,5
TRICINE	N-[Tris-(hydroxymethyl)-methyl]-glycine	55185	179,2	7,4-8,8	8,15
TRIS	2-Amino-2-(hydroxymethyl)-propane-1,3-diol	55065	121,1	7,2-9,0	8,30

Buffer Substance	Chemical name	Sinus code	M.W.	Buffering range	pka (20°C)
ACES	N-(2-Acetamido)-2-amino-ethane sulfonic acid	55001	182.2	7.1–7.5	6.90
AMPD	2-Amino-2-methyl-1,3-propandiol	55103	105,1	7,8–9,7	n.a.
AMPSO	N-(1,1-Dimethyl-2-hydroxyethyl)-3-amino-2-hydroxy-propanesulfonic acid	55104	227,3	8,3–9,7	n.a.
BES	N,N-Bis-(2-hydroxyethyl)-2-aminoethanesulfonic acid	55106	213,3	6,4–7,8	7,15
BICINE	N,N-Bis-(2-hydroxyethyl)-glycine	55105	163,2	7,6–9,0	8,35
BISTRIS	Bis-(2-hydroxyethyl)-imino-tris-(hydroxymethyl)-methane	55107	209,2	5,8–7,2	6,4
CHES	2-(N-Cyclohexylamino)-ethane sulfonic acid	55109	207,3	8,6–10,0	9,55

To get a better sense of the utility of buffer solutions, let's compare the difference between adding a strong base to a buffered solution, and adding the same amount of strong base to water. This is done in the examples below.

EXAMPLE

If 10 mL of 1M NaOH are added to one litre of a buffer that is 0.3 M acetic acid and 0.2 M sodium acetate ($Na+CH_3COO^-$), how much does the pH change? The pK_a for acetic acid is 4.76.

Answer: In the original solution, acetic acid is the weak acid, and acetate is the conjugate base. Thus, [base] = 0.2 M, and [acid] = 0.3 M. Using the Henderson-Hasselbalch equation,

$$pH = pK_a + \log\frac{[Base]}{[Acid]}$$

H = 4.72 + log (0.2/0.3)

pH = 4.54

NaOH is a strong base, and dissociates completely. Therefore, adding the strong base results in (0.01 litre)(1 M) = 0.01 moles of OH- ions into the solution. *All of the OH– ions will react with the acetic acid to form acetate ions*:

$$CH_3COOH + OH^- \rightleftharpoons CH_3COO^- + H_2O$$

Thus, when NaOH is added, the number of acetic acid molecules in the solution will decrease, but the number of acetate molecules in the solution will increase. In this case, 0.01 moles of acetic acid will be used up to neutralize the 0.01 moles of OH-, and this will form 0.01 moles of acetate ion in the solution.

At the end of the reaction:

Moles of acetic acid	=	**Moles of acetic acid before the reaction**	–	**Moles of acetic acid used up in the reaction**		
	=	0.3 moles	–	0.01 moles	=	0.29 moles
Moles of acetate	=	moles of acetate before the reaction	+	moles of acetate made in the reaction		
	=	0.2 moles	+	0.01 moles	=	0.21 moles

Now we are almost ready to plug back into the Henderson-Hasselbalch equation to find the new pH of the solution. We just need to remember that the volume has changed slightly, from 1 litre (1000 mL) to 1 litre plus 10 mL (1010 mL). The Henderson-Hasselbalch equation requires that the [Base] and [Acid] be molar (M, or mol/L) quantities. However, notice how that because both the acid and base species are in the same volume, the volume correction cancels out:

$$pH = pK_a + \log\frac{[Base]}{[Acid]}$$

pH = 4.72 + log 1.38

pH = 4.86

The pH change is 4.86 – 4.54 = 0.32 units.

Example: How much does the pH change when 10 mL of 1 M NaOH is added to 1 litre of water?

Answer: Pure water has a pH of 7.0. NaOH is a strong base, and once it has been added to the water it will dissociate completely, contributing 0.01 moles OH- ions, in a total volume of 1010 mL (1 litre plus the 10 mL NaOH added).

Thus,

$[OH^-]$ = 0.01 moles/ 1.01 litre = 0.0099 M

Plugging into the Kw expression: $[OH^-][H3O+] = 10–14$:

$(0.0099)[H_3O+] = 10–14$

$[H_3O+] = 1.01 \times 10–12$ M

Then plugging into the pH equation:

$pH = –log [H^+]$

$pH = –log 1.01 \times 10^{-12}$

$pH = 11.99$

The pH has shifted from the neutral pH of 7 for the pure water to a pH of almost 12 with the NaOH added, a shift of 5 pH units!

STABILIZING pH OF AQUEOUS SOLUTIONS

In example, the pH of the buffer solution changed by less than 0.5 units when the NaOH was added. Yet adding the same amount of NaOH to pure water resulted in a 5 unit pH shift in example . Clearly, the conjugate acid/base pair was able to protect the solution from a large change in pH.

A buffer is a solution that resists a change in pH when acids or bases are added to it. How is this possible? Buffers work by acting a little bit like a sponge, soaking up excess H3O+ or OH– ions when they are added to a solution. The dissociation reaction for acetic acid shows that a solution of acetic acid contains acetate ions:

$$CH_3COOH + H_2O \rightleftharpoons CH_3COO– + H_3O^+$$

How do acetic acid and the acetate ion work to buffer the solution? If a strong acid, such as hydrochloric acid, is added to this buffer, the H3O+ ions generated will react with the acetate ions, removing them from the solution:

$$CH_3COO^- + H_3O+ \rightleftharpoons CH_3COOH + H_2O$$

acetate ion, hydronium ion, acetic acid, water

Similarly, if a strong base, generating lots of OH^- ions, were added to the acetic acid buffer solution.

$$CH_3COOH+ HO– \rightleftharpoons CH_3COO – H_2O$$

acetate acid, hydroxide ion, acetate acid, water

The OH- ions are removed via a reaction with acetic acid. It is important to remember that buffers cannot maintain their pH indefinitely as more acid or base is added.

Imagine slowly pouring a bucketful of water onto a small sponge on the kitchen floor. At first, as the water pours out of the bucket it is absorbed by the sponge, keeping the kitchen floor dry.

But once the sponge is soaked through, it can hold no more and the water spills onto the floor as fast as if the sponge weren't there.

In the same way, buffers will protect the pH of the solution to some extent, but if they are inundated by large amounts of H_3O+ or OH– ions, all the available conjugate acid or base molecules will have been used up and pH changes will rapidly occur.

Thus, buffers only work as long as the amount of conjugate acid and base ions are large compared to the H_3O+amount of or OH– ions to be removed.

Remember from the Henderson-Hasselbalch equation that the numbers of conjugate acid molecules is equal to the number of conjugate base molecules when the pH = pKa:

$pH = pK_a + \log ([Base]/ [Acid])$

if [Base] = [Acid], then

$pH = pK_a + \log 1$

$pH = pK_a$

Because the conjugate acid and conjugate base molecules are available in equal amounts when the pH = pKa, the buffer solution has the strongest ability to protect against pH changes caused by incoming H_3O^+ or OH-ions when the pH of the solution is close to the pK_a of the conjugate acid.

If the pH of the solution strays too far from the pKa point, the buffer uses up all its available conjugate acid or base molecules, and the pH starts to fall dramatically.

At this point, the buffering capacity of the buffer has been exceeded. this can be seen in the buffer titration graph below. Buffers protect the pH of a solution best within one pH unit of the pKa.

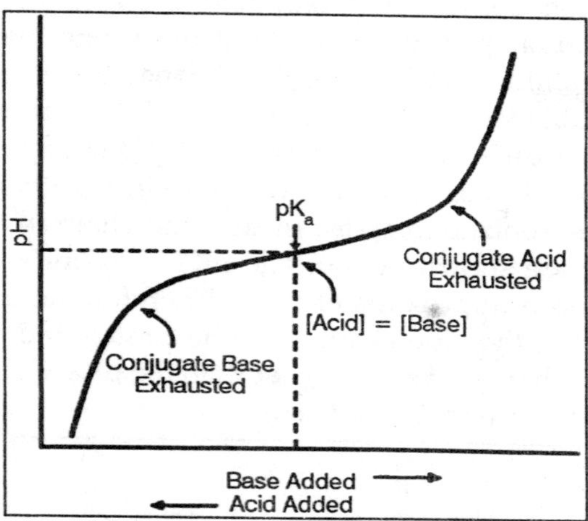

Fig. Buffer Titration: Within One Unit of the pKa, the Solution Resists Changes in pH.

Buffers are important for biochemistry because the structures (and therefore the functions) of biological molecules are stable within a relatively narrow range of pH values.The evolution of life, which is believed to have begun in water, was likely due in part to the stable pH of seawater. Even today, the largest buffered systems in the world are the Earth's oceans. Carbon dioxide (CO2) from the Earth's atmosphere reacts with H2O to produce carbonic acid, H2CO3. Carbonic acid in turn reacts with water to form bicarbonate:

$$H_3CO_3 + H_2O \rightleftharpoons HCO_3 - H_3O^+$$

crabonic acid | water | bicarbonate ion | Hydronium ion

Furthermore, the bicarbonate ion can also react with water to form carbonate:

$$H_3CO_3^- + H_2O \rightleftharpoons HCO_3 - H_3O^+$$

bicarbonate ion | water | Carbonate ion | hydronium ion

The carbonate and bicarbonate ions act together as conjugate acid/base pair to keep the pH of the ocean at about 8.2. In the living organisms, proteins (the molecules which do most of the"work" of a cell) are very sensitive to acid concentration, because they only fold into their proper three-dimensional shapes within a small pH range. Some of the individual amino acid building blocks that make up a protein have ionizable side chains that change their ionization state, or charge, at different pH. Thus, if the pH changes, some of the attractive forces that hold the protein together will also change. The ionization states of the side chain of the amino acid lysine are shown below. The pKa of lysine's side chain amino group is approximately 10.8. Therefore, at a pH lower than the pKa, the side chain is in the protonated (conjugate acid) state; whereas at a pH higher than the pKa, the side chain is unprotonated (conjugate base).

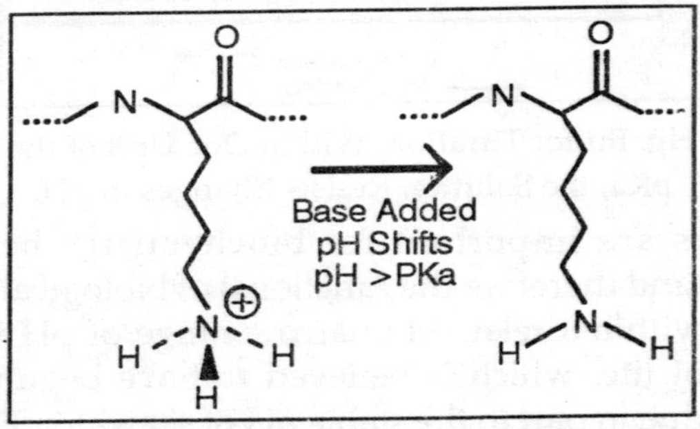

Fig. Amino Acid Charge changes.by Environment.

Proteins consist of long chains of amino acids, many of which include acidic or basic groups. These groups are chaged at physiological pH, and form ionic interactions with each other that contribute to the stability of the folded protein. A change in pH of the protein's aqueous environment may alter the charge of some of these charged groups, disrupting the ionic interactions and the stability of the protein's structure.

Remember that the Henderson-Hasselbalch equation was used for calculating the pH of a solution containing a weak acid and its conjugate base. Since such a solution is a buffer, the Henderson-Hasselbalch equation is extremely useful for calculations involving buffers, allowing the following:

- Determination of the pH of buffered solutions.
- Determination of the ratio of conjugate base to conjugate acid at a given pH.

Example: A solution of lactic acid has twice as many conjugate acid molecules as conjugate base molecules. If the pKa of lactic acid is 3.85, what is the pH of such a solution?

***Answer*:**

We are given that the concentration of conjugate acid is twice the concentration of conjugate base.

Thus,

[acid] = 2 [base]

Substituting into the Henderson-Hasselbalch equation:

pH = pK_a + log ([base]/[acid])

pH = 3.85 + log ([base]/2[base])

pH = 3.85 + log (0.5)

pH = 3.85 + (–0.30)

pH = 3.55

Question: If a buffer solution contains an acid as one component and a base as the other, can those two componenets be HCL and NaOH?

Answer: Your assumption is correct. A buffer is made of a weak acid and its conjugate base or a weak base and its conjugate acid. So neither HCl nor NaOH will be a component in a buffer; they are both strong. An example of a buffer is acetic acid (a weak acid) and sodium acetate (a conjugate base of acetic acid as a salt).

Incidentally, in the example you gave, HCl in water dissociates to hydronium ions and chloride ions. NaOH in water dissociates to hydroxide ions and sodium ions. When the two are mixed, the hydronium ions and the hydroxide ions combine to form water leaving behind the sodium and chloride ions. If you have the same amount and concentration of HCl as NaOH you will end up with salt water because sodium chloride is simply table salt.

BUFFERS AND IONIZATION OF AMINO ACIDS

- Buffers resist change. pH buffers resist change in pH when acid (H+) or base (OH–) is added.
- pH = –log [H+] –where bracket indicates molar

concentration. If [H+] = 10^–7 M, then pH = 7. This is neutral pH since [H+] = [OH–] and it is middle of pH scale. Draw a pH scale by calculating the pH of 1 M HCl and 1 M NaOH. Use the equation Kw = 10^–14 = [H+] × [OH–], to find the pH of 1 M NaOH.

- pH of a buffered solution can be found by using the Henderson-Hasselbach equation:

pH = pK + log([A–]/[HA])

All acids and bases have a conjugate acid (HA) and a conjugate base (A–). At [HA] = [A–], pH = pK (prove this with the above equation).

At the pK, a buffer maximally resists change in pH and...the buffering zone is considered to be 1 pH unit above and below the pK.

CONCEPTS

Buffering zone = pK +/–1 pH unit:

- At pH = pK, [HA] = [A–].
- At pH below pK, [HA] > [A–].
- At pH above pK, [HA] <[A–].
- Amino acids (AAs) are more complex than simple buffers and...have at least two pK values because AAs have at least two ionizing groups:

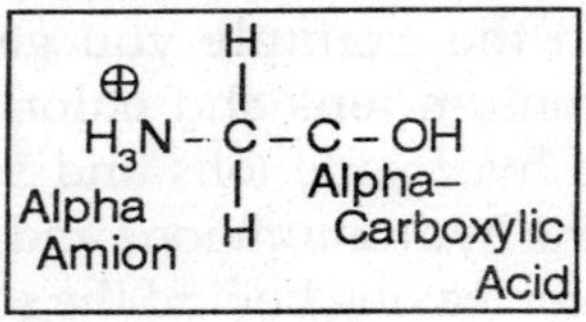

$$\overset{\oplus}{H_3N}-\underset{H}{\overset{H}{C}}-C-OH$$

Alpha Amion

Alpha–Carboxylic Acid

Glycine Structure

Example: Glycine has an alpha-amino group and alpha-carboxylic acid group. Each ionizable group has a pK (called pK–1 and pK–2). Glycine pK–1 = 2.3 (alpha-carboxylic acid); pK–2 = 9.6 (alpha amine).

Draw a titration curve for Glycine for practice. In this class, we treat all ionizing groups with in a molecule, as acting independently of other ionizing groups. The pH on the titration curve where an AA has no net charge is called the pI or isoelectric point. The pI is calculated by averaging the two pK values on either side of the neutral form (ie form with no net charge).

All AAs must have at least 3 ionic forms:

- "AA+1","AA0", and"AA-1".
- Where"AA+1" = AA form with net charge of +1,
- "AA0" = neutral form with no net charge, and
- "AA-1" = AA form with net charge of –1.

While an AA may have other ionic forms, only these 3 forms count when finding pI:

- AA+1 goes to AA0 via pK–below.
- AA0 goes to AA-1 via pK–above.

Idea

pI = average of (pK-below) + (pK–above).
where pK-below is the pK between AA+1 and AA0.
and pK-above is the pK between AA0 and AA–1.

For more complex AAs, like Asp, Glu, Lys, Arg and His - with 3 ionizing groups, Other forms exist with charges other than +1, zero, and -1. For these AAs, the best approach is to start with the fully protonated form (ie. One with all the protons it can take on its ionizable groups).

Calculate the net charge on this group (it must be positively charged), then titrate the AA to remove the first proton and find net charge again. Continue doing this until all the protons have been removed. Keep track of this by making a simple model. Then, find the form with no net charge (ie. AA0) and use the pK values, which govern the transition from AA+1 to AA0 and AA0 to AA-1, as the ones to calculate the pI.

His has 3 ionizing groups, alpha-carboxylic acid (pK 1.8), side-chain amino (pK 6.0) and alpha-amino (pK 9.2):

- His+2 goes to His+1 via pK 1.8;
- His+1 goes to His0 via pK 6.0;
- His0 goes to His-1 via pK 9.2.

Therefore, pI = (6.0 + 9.2)/2 = 7.6:

- Some of you have trouble figuring out which group ionizes first. Here are some rules to apply to help you.

- Carboxylic acids ionize at acidic pH; ie. Carboxylic acids give up their protons at acid pHs.
- Amino groups ionize at basic pH; ie. Amines give up their protons at basic or alkaline pHs.
- When groups with a similar chemical nature are present:
  (a) Carboxylic acids near an amino group have a more acidic pK than isolated carboxylic acids.
  (b) Amino groups near a carboxylic acid have a more acidic pK than isolated amines.
  (b) Aromatic amines (like in His side chain) have a pK near neutral.

Apply the above rules and concepts to calculate the pI for all the other complex AAs.

Asp, Glu, Lys, and Arg. Their pK values and structures are shown below:

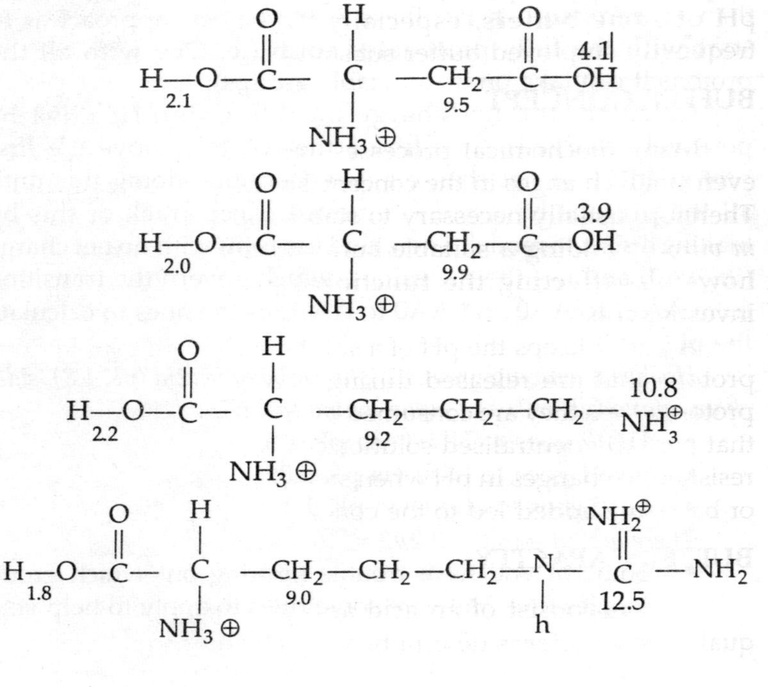

```
         O      H           O
         ||     |           ||   4.1|
   H—O—C—    C    —CH2—C—OH
     2.1        |       9.5
               NH3 ⊕

         O      H           O
         ||     |           ||   3.9
   H—O—C—    C    —CH2—C—OH
     2.0        |       9.9
               NH3 ⊕

         O      H
         ||     |                        10.8
   H—O—C—    C    —CH2—CH2—CH2—NH3⊕
     2.2        |       9.2
               NH3 ⊕

      O      H                               NH2⊕
      ||     |                                ||
H—O—C—    C    —CH2—CH2—CH2—N—    C   —NH2
 1.8         |        9.0              |     12.5
            NH3 ⊕                      h
```

BIOLOGICAL BUFFERS

Many biochemical processes are markedly impaired by even small changes in the concentrations of free H+ ions. It is therefore usually necessary to stabilise the H+ concentration *in vitro* by adding a suitable buffer to the medium, without, however, affecting the functioning of the system under investigation. A buffer keeps the pH of a solution constant by taking up protons that are released during reactions, or by releasing protons when they are consumed by reactions. The observation that partially neutralised solutions of weak acids or bases are resistant to changes in pH when small amounts of strong acids or bases are added led to the concept of the'buffer'.

The Biological Buffers imparts basic knowledge on biological buffer substances and support in terms of selecting the best buffer substance for your assay system. Be aware of the fact that some buffers serve as substrate/ inhibitor of enzymes (e.g. phosphate) or the temperature sensitivity of the pH of some buffers, especially Tris, probably the most frequently employed buffer substance.

BUFFER CONCEPT

Many biochemical processes are markedly impaired by even small ch anges in the concentrations of free H+ ions. It is Therefore usually necessary to stabilise the H+ concentration *in vitro* by adding a suitable buffer to the medium, without, however, affecting the functioning of the system under investigation.

A buffer keeps the pH of a solution constant by taking up protons that are released during reactions, or by releasing protons when they are consumed by reactions. The observation that partially neutralised solutions of weak acids or bases are resistant to changes in pH when small amounts of strong acids or bases are added led to the concept of the'buffer'.

BUFFER CAPACITY

Buffers consist of an acid and its conjugated base. The quality of a buffer is determined by its buffer capacity, i.e. its resistance to changes in pH when strong acids or bases are added. In other words: the buffer capacity corresponds to the amount of H+ or OH- ions that can be neutralised by the buffer. The buffer capacity is related to the buffer concentration. The graph described by the relation of the pH to the addition of H+/OH– ions is called the titration curve.

The point of inflection of the curve corresponds to the pKa value. At this point, the buffer capacity is at its maximum at the pKa value. This point therefore corresponds to the mid-point of the pH range covered by the buffer and is where the concentration of acid and base is the same. In the area of this pH range, therefore, relatively large amounts of H+/OH– ions result in only small changes in pH. A basic principle is that a buffer that has a pH value of one pH unit above or below the pKa value loses so much buffer capacity that it no longer has any real buffer function.

Based on the Henderson-Hasselbalch equation pH = pKa + log [A–]/[HA] for the calculation of the pH of a weak acid or alkaline solution, the ions in the water must also be taken into account when working in pH ranges below 3.0 and above 11.0. Most biochemical reactions, however, take place in the pH range between 6.0 and 10.0.

THE PH VALUE

Conductivity can also be detected in highly purified water due to the OH– and H3O+ ions resulting from the autoprotolysis of water. This intrinsic dissociation of the water is an equilibrium reaction and the product of the concentrations of the two ions represents a constant:

$$K = [H3O+] \times [OH–].$$

This value is depends on the temperature only and is 10-14 for purified water at 22°C. Depending on which of the two ions is present in a higher concentration in a solution, the solution is called to be acidic or alkaline. To express this fact in terms of a simple number, the negative exponent of the easily measurable hydronium ion concentration [H3O+] was chosen.

This dimensionless number is called the pH value. The

pH can also be described as the negative, decadic logarithm of the hydronium ion concentration of a solution:

$$pH = -\log [H_3O^+]$$

The hydronium ion concentration of pure water is 10–7 mol/L, as can be derived from the above equation for the ion product. Therefore, the pH value is 7. In an acidic solution, the concentration of H_3O^+ ions is increased (e.g. from 10–7 mol/L to 10–2 mol/L) and the pH is therefore < 7. In an alkaline solution, the hydronium ion concentration is decreased and this results in a pH > 7.

THE PKA VALUE

To change the pH value of a solution, substances are dissolved in the solution which release H^+ ions into the water (acids), thus raising the H_3O^+ ion concentration and lowering the pH value, or which decrease the H_3O^+ ion concentration and thus increase the pH value by taking up H^+ ions (bases). As always in chemistry, this reaction is an equilibrium reaction, and the capacity of a substance to shift this equilibrium in one direction or another is determined by the potency of the acid. It is calculated from the equilibrium constants using the following equation $K = [A] \times [H_3O^+]/ [HA]$, requirements.

DILUTION AND RINSE BUFFERS

Rowmanowsky stains (Giemsa, Leishmans, Wrights etc) are pH sensitive. Tinctorial (stain picture) quality can be altered by changes in the final pH of staining reagent, this can be caused by differences in the quality or source of the water used to dilute such stains. Tinctorial quality and consistency can be maintained by the use of dilution buffers, which when added to the diluent water will produce stain reagents of the required pH.

Acid pH (<7.0) will emphasize eosinophilic staining.

Basic pH (>7.0) will emphasize azureophilic (basophilic) staining.

In general, blood stains are diluted to produce solutions of slightly acid pH, 6.50-6.80.

We offer a range of buffer concentrates covering pH's 6.50/

6.80/ 7.00/ 7.20/ 7.40/ 7.60, supplied in 10 vial or 25 vial packs each vial makes 5 litres of dilution buffer. They offer great value for money and the certainty of preparing blood stains of the required pH time and again.

To maintain the tinctorial qualities of the final stain slides should be rinsed in the same buffer.

Concentrates - ready-to-use.

High quality, accurate, time saving and good value for money.

Chapter 3

Water

DEFINITION

Water has many unique properties that make it essential to all life. Most of water's unique properties are a result of the hydrogen bonding between water molecules.

Water is an excellent solvent. When ionic compounds are placed into water, the ions dissociate or separate. Polar covalent compounds, because they too have charged poles, also dissolve in water. Nonpolar covalentcompounds, however, do not dissolve in water. Thus polar covalent compounds are hydrophilic (water loving) while nonpolar covalent compounds are hydrophobic (water fearing).

Water has high cohesion. Individual water molecules tend to"stick" with other water molecules due to hydrogen bonding. This leads to two characteristics of water: a high surface tension and strong capillary action. Water's high surface tension is what allows some insects to run across the surface of the water. It is also the reason why belly-flops hurt so much. Water's strong capillary action is what allows the liquid level in a straw to be higher than that in the surrounding drink. Using hydrogen bonding, water molecules attract others up the sides of the straw. This effect is the more noticeable in thinner straws.

You're right if you think it's terribly difficult to select an aspect or two to discuss when the larger topic is water. There's all sorts of fascinating stuff! Everything from the role of the waters of hydration in antibody-binding to the effect of solvation for stabilizing protein structure to the dependence of water's dielectric and pH on temperature and pressure. This last topic is particularly exciting as we learn more about the thermophiles living at deep sea vents. And to paraphrase a guest I heard on a radio science-talk show recently, once we better understand the life living in our oceans, we'll be much better prepared to understand possible life on other planets.

But most importantly, because the aqueous environment influences all of the chemistry that takes place in it, when you are discussing water you get to address just about anything, while still ostensibly focusing on the substance itself.

Before we get to water at extremes of temperature and pressure, let's touch briefly on the oh so neglected role of water in our everyday life. And just to be sure we start out on a provocative note, let me here quote Felix Franks who told me

BIOCHEMISTRY AS CHEMISTRY OF WATER

We immediately see the evident truth in this statement when we consider the complex thermodynamics of protein and peptide structural stability and the prominent role of solvent in influencing this equilibrium. Proteins unfold from their native fold upon the addition of urea or other denaturants, peptide helicity increases upon addition of TFE, and alcohol is even known to induce helical structure in regions that otherwise adopt beta-sheet conformations.

While the helix inducing effects of alcohol upon peptide conformation have long been used to probe and enhance latent structure, thereby making it measurable by CD or NMR, a recent paper illustrates how this dependence of structure on solvent conditions may be harnessed to make a sort of'conformational switch'. Successful molecular design or peptide engineering applications will necessarily include an assessment of solvation effects for conformational stability.

Turning from a macroscopic picture of the role of water and aqueous solutions in establishing the equilibria of proteinaceous substances to a more microscopic view, we can consider the surprising role played by discrete water molecules in antigen-antibody recognition. In a 1993 paper in Current Opinion in Immunology, Mariuzza and Poljak report a case in which interfacial water plays a critical role in determining the complementarity of the antigen and antibody surfaces.

This work was illuminating, since previous work was leading to the conclusion that the packing density was lower at protein interfaces than in protein interiors. It also appeared that polar atoms at these interfaces were apparently tolerating the absence of neutralizing partners - far more than they do in protein interiors. In this work, a 1.8 angstrom crystal structure of an immune complex was studied and revealed that the water network was an integral part of the interface.

As pointed out by these authors, the analysis of low resolution crystal structures may in fact lead to erroneous conclusions on the nature of the interface in these complexes. Furthermore, the evidence reported in this work points to the inescapable conclusion that the intermolecular complementarity essential for molecular recognition is not necessarily imparted solely by the large molecules.

And, contrary to what our notions of solvent entropy might lead us to expect, bound water may play a role in stabilizing the structure of single proteins as well as that of complexes. And, this bound water may, apparently, serve this function at the surface of the protein, and not solely while confined to the interior. That's just the tip of the iceberg as far as far as water and its impact for biological systems at what is considered physiological temperature and pressure; the realm normally explored by biochemists and pharmacologists.

Much, however is being learned about the so-called"hot life", life found at thermal vents deep in the ocean floor. Reports have documented the existence of organisms at temperatures in excess of 110 deg. C (see Shock for a list of references). These organisms are referred to collectively as thermophiles. It is important to note the enormous temperature variation of sea water at these depths, with a documented range from 2 deg. C to 900 deg. C at depths of a few km where the high pressure precludes boiling. Hot life, or the existence of thermophiles, has been noted at temperature in excess of 100 deg. C at these elevated pressures, and it remains a matter of some dispute whether or not bacteria can exist at even higher temperatures still.

So, what are the properies of ambient sea water for organisms at deep sea thermal vents? We can expect these properties to influence the kinetics and thermodynamics of the biochemistry pertinent to this life, just as they influence the systems with which we are, perhaps, more familiar. We can begin to understand the fluid environment presented to thermophiles by considering the temperature dependence of a few properties of pure water.

A quick look at the following data from Shock show the dramatic effect of temperature on water's dielectric. Water at 25 deg. C and 1 atm pressure has the familiar dielectric near 80; however near water's critical point the dielectric has fallen to less than 10. For comparison, ethanol has a dielectric of roughly 25 at STP.

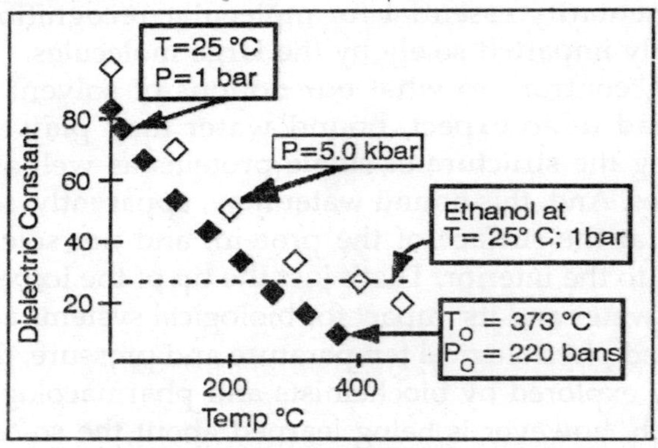

Fig. From Shock.

In this graph all white and blue diamonds pertain to water data points. The blue diamonds are for pressures below the critical point where water exists in two phases, while the white diamonds pertain to 5 kbar data. For the low pressure data, two key data points are indicated. The grey dashed line indicates the dielectric of ethanol at standard conditions. (This graph is drawn to approximate that shown in the original paper which should be consulted for quantitative work.)

The weak pressure dependence of this data should be noted. In order to understand the relevance of this sort of analysis to a complete understanding of ambient conditions at deep sea thermal vents, it should be understood that the pressure at these depths ranges from a few hundred bars to several kilobars. Shock should be consulted for additional details, data and a general groovy time.

Well, if all this is true, then water must have a pretty different structure at these T&P's than we are used to. Right? Not surprisingly, this is what is being found. Some of these results are shown in this graph:

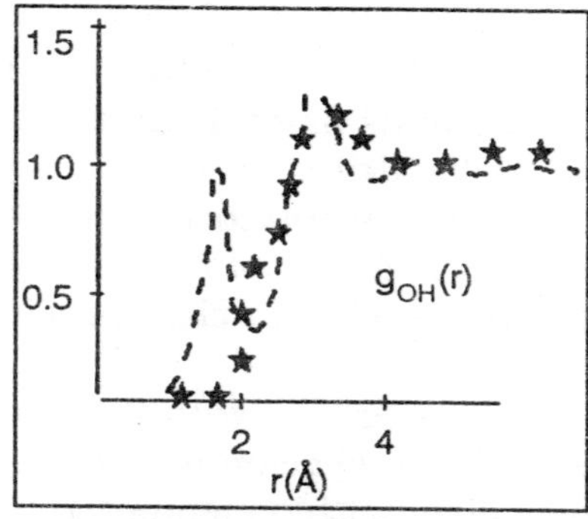

Fig. From Postorino

The pair distribution function for intermolecular OH atom pairs is shown as a function of interatomic distance for water at 25 deg. C (red dashed line) and 400 deg. C. (This graph is drawn to approximate that shown in the original paper which should be consulted for quantitative work.)

This particular pair distribution function tells us the likelihood with which a H atom will be found at any distance from an O atom (The intramolecular data isn't shown here, but the water molecule does have the same structure at 400 deg. C as at 25 deg. C.

This graph considers only intermolecular associations.) Immediately, then, we see that at room temperature it is relatively likely to find these atom pairs separated by 1.95 angstroms. This is the hydrogen bond. And then there are strong secondary structural phenomena, manifest as the damped oscillations that we see here out to beyond 5 angstroms, all due to the overwhelming hydrogen bonding Behaviour.

Looking at the high temperature data, it's clear that the hydrogen bonding peak, which dominates the room temperature data, is completely gone.

The OO correlation function, not reproduced here, shows a slight broadening of the nearest-Neighbour peak with temperature.

The high temperature OO pair correlation function also yields a larger coordination number than is seen at room temperature.

In other words, each water molecule is surrounded by more water molecules in its first solvation shell at 400 deg. C than at 25 deg. C. At 400 deg.

C the packing is not restricted by the severe hydrogen-bonding geometrical constraints of the the room temperature fluid. This result is in line with the close packing Behaviour of most liquids.

These results taken together with those for the dielectric phenomena described above begin to paint a picture of high temperature water. Water at high temperature exhibits many of the properties we associate with less polar liquids under standard conditions.

In order to understand something of the biochemistry that might pertain under extreme aqueous conditions, Shock has been estimating the stability of peptidic species at elevated temperatures and pressures.

Specifically he has been asking questions like"How does the free energy of the peptide bond change with temperature and pressure?" A snapshot view of some of this work will serve to further the discussion of aqueous properties under extreme conditions while at the same time giving a hint for some of the very different biochemistry that will undoubtedly be revealed by further study in this area.

Considering only Glycine here, the free energy of interest here is that which determines the following equilibrium:

$$\text{Glycine} + \text{Glycine} = \text{Glycylglycine} + H_2O$$

In other words, what is the free energy cost to form the peptide bond between two Glycine molecules? This equilibrium can be calculated from the free energies of formation for the individual species involved in this reaction, Glycine and Glycylglycine, and the temperature dependence of this free energy in the following way:

$$\Delta G^{\circ}_{PT} = \Delta G^{\circ}_{f} - S_{Pr,Tr}(T - Tr) + \int C^{\circ}_{P}$$

$$-T\int C^{\circ}_{P} d\ln T + \int V^{\circ} dP$$

This equation has the usual host of characters. G represents free energy, S is entropy, Cp is heat capacity, V is volume and P is pressure. If any of this is murky please consult the original paper by Shock.

What we're basically doing is getting the free energy (OK, so it's really the standard partial molal Gibbs free energy) at any T&P from that at STP by integrating over the temperature and pressure dependence of the heat capacity and volume.

In order to do this, we need to know something about the way heat capacity and volume depend on the thermodynamic variables of T and P, respectively. We'll look at this first:

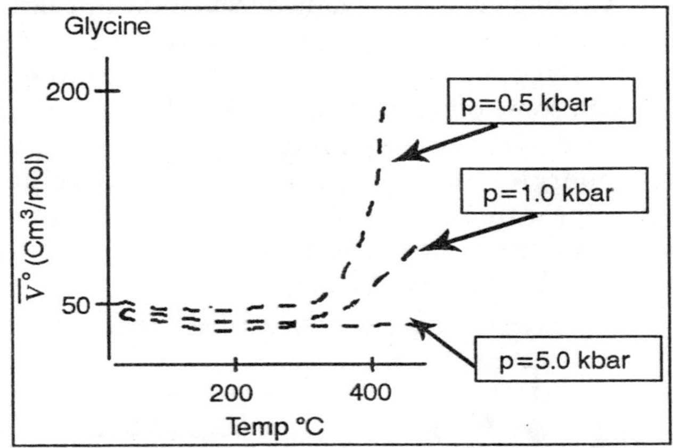

Standard partial molal volume for Glycine. From Shock. (This graph is drawn to approximate that shown in the original paper which should be consulted for quantitative work.)

These results show the very large dependence of the molal volume on temperature at lower pressures, and the much weaker temperature dependence at extremely high pressures. Here are the results for the heat capacity:

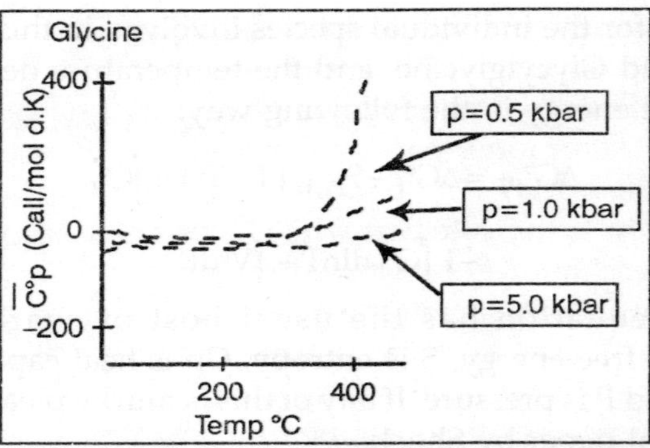

Standard partial molal heat capacity (constant pressure variety) for Glycine. From Shock. (This graph is drawn to approximate that shown in the original paper which should be consulted for quantitative work.) Pretty much the same thing. A large T dependence is witnessed at lower temperature, but this Behaviour is convincingly squashed at high pressure.

The results for Glycine have been shown here. Analogous results for Glycylglycine are obtained from these by calculating a regression equation that relates, for instance, the partial molal volume of Glycine to that of Glycylglycine. Putting it all together with some other necessary data (like the free energy of formation at STP, and the entropy at STP; you got it, Shock) the free energies of formation for both Glycine and Glycylglycine can be obtained at any temperature and pressure, and from these the free energy governing the equilibria may be directly obtained.

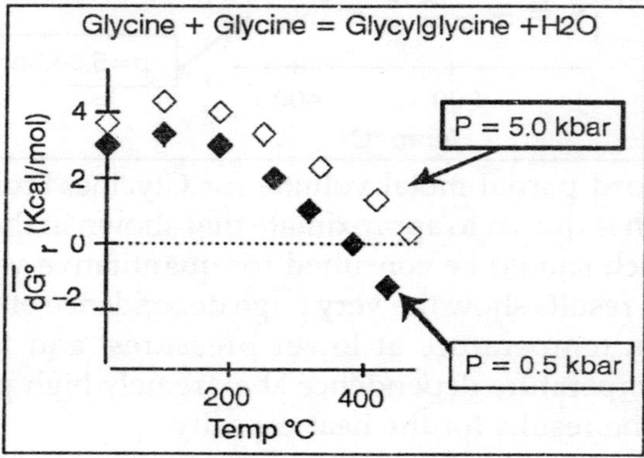

Free energy required to form the Glycylglycine dipeptide from the constituent Glycine moieties.

From Shock. (This graph is drawn to approximate that shown in the original paper which should be consulted for quantitative work.)

The results indicate that while it costs free energy to form the dipeptide at lower temperatures and pressures (at STP the free energy of this reaction is roughly 3.5 kcal/mol), at extremes of temperature and pressure the dipeptide is the thermodynamically favoured species.

In any event, these results are consistent with those for many other dipeptides not covered here and predict that peptide bonds become more stable at elevated temperatures. This Behaviour just begins to hint at the interesting biochemical phenomena that will be revealed for life in water under extreme conditions where water has very different properties from those to which we are accustomed.

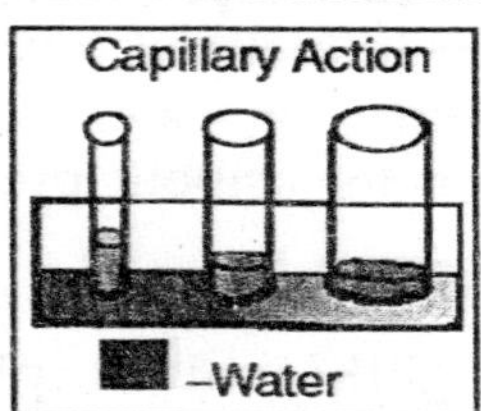

Water has a high specific heat capacity. It takes a lot of energy to change the temperature of water. This characteristic makes water a good insulator and a good coolant.

When you sweat, your body is using water as a coolant. The eVapourating water removes heat with it. At night, oceans are a good insulator. The energy that the sun spent in heating the water all day is slowly released into the night.

Finally, water is the most dense at 4 degrees Celsius. Hydrogen bonding arranges water molecules into hollow"cells" when water freezes, making it less dense than liquid water.

This characteristic of water is what allows fish in lakes and ponds to survive in the winter. When the water freezes, it becomes less dense and floats to the surface -- leaving the bottom of the lake or pond unfrozen.

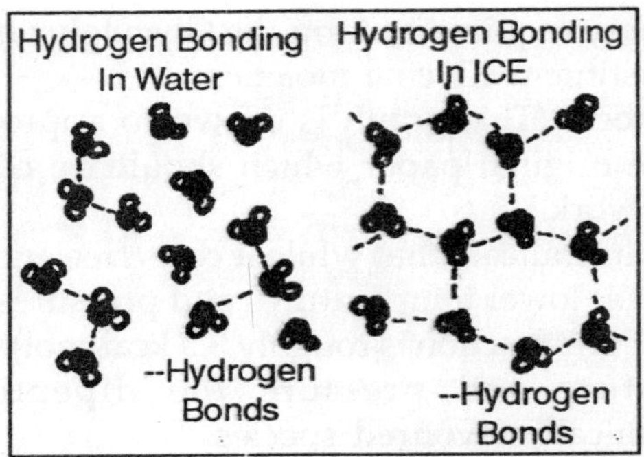

WEAK INTERACTIONS IN AQUEOUS SYSTEMS

Water is a good solvent for polar solutes with which it forms hydrogen bonds. Non-polar compounds dissolve poorly in water. They can not form a hydrogen bond with the solvent. Numerous weak non covalent interactions influence the folding of macromolecules. The physical properties of aqueous solutions are strongly influenced by the concentration of solutes. Ex: The tendency for water to move across a semi permeable membrane is the osmotic pressure.

IONIZATION OF WATER

Pure water is slightly ionized. pH is defined as the negative logarithm of the molar hydronium ion concentration. An acid that ionizes completely in water is a strong acid. An acid which is partially ionized in water is a weak acid. pk provides the measure of the ease of the dissociation of the proton. The lower the number the stronger is the acid.

BUFFERING AGAINST PH

Buffers are aqueous systems that tend to resist in changes in pH when small amounts or acids or base are added. It consists of a weak acid and its conjugate base. As an example a mixture of equal concentrations of acetic acid and acetate ion is a buffer system. Every life form is extremely sensitive to slight pH changes. Human blood for example needs to remain within the pH range of 7.38 to 7.42. The relationship between pH and pk is described by Henderson - Hasselbach equation. It is used to calculate the pH of biological fluids. Two especially important biological buffers are the phosphate and bicarbonate systems. The property of a solution that depends on the concentration of a dissolved solute but not on its chemical identity is known as colligative properties. Substances that do not ionize in water are non electrolytes. Vapour pressure, boiling point, freezing point and osmosis are colligative properties.

WATER AS A REACTANT

Water is very often a direct participant in the chemical reactions of living cells. Formation of ATP from ADP and inorganic phosphate is an example of a condensation reaction in which the elements of water are eliminated. Two amino acid molecules are covalently joined through a peptide bond with a loss of water molecule.

FITNESS OF AQUEOUS ENVIRONMENT FOR LIVING ORGANISMS

The high specific heat of water is useful to cells and organisms because it allows water to act as a'heat buffer', keeping the temperature of an organism relatively constant as the temperature of the surroundings fluctuates.

SUMMARY

Water is a good solvent for polar solutes with which it forms hydrogen bonds. Non-polar compounds dissolve poorly in water. They can not form a hydrogen bond with the solvent. Numerous weak non covalent interactions influence the folding of macromolecules.

The physical properties of aqueous solutions are strongly influenced by the concentration of solutes. The equilibrium constant describes whether an acid or a base is strong or weak. A mixture of weak acid and its salt resists changes in pH caused by the addition of H+ or OH-. The mixture functions as a buffer.

The relationship between pH and pk is described by Henderson-Hasselbach equation. It is used to calculate the pH of biological fluids. Water is very often a direct participant in the chemical reactions of living cells.

PROPERTIES AND BIOMOLECULAR STRUCTURE

Water is necessary for life. Many plant and animal adaptations conserve water - the thick skin of desert cacti and the intricate structure of the mammalian kidney are just two examples.

Planetary scientists look for evidence of liquid water when speculating about the possibility of life on other planets such as Mars or Jupiter's moon, Titan.

WATER REMARKABLE PROPERTIES

High Surface Tension

Despite being denser than water, small objects, such as aquatic insects, can stay on top of water surface.

High Boiling Point

Relative to its molecular weight, water boils at a high temperature. For example, ammonia, with a molecular weight of almost 17, boils at –33° C, while water, with a molecular weight of 18, boils at 100° C.

Density is Dependent on Temperature

Solid water (ice) is less dense than liquid water. This property means that lakes and ponds freeze from the top down, a benefit to the fish living there, who can overwinter without being frozen solid.

PROPERTIES OF WATER AND HYDROGEN BONDS

Water has a dipole, that is, a separation of partial electrical charge along the molecule. Two of oxygen's six outer-shell electrons form covalent bonds with the hydrogen. The other four electrons are nonbonding and form two pairs. These pairs are a focus of the partial negative charge, and the hydrogen atoms ..ıd negative charges attract each other, so that the Po.. and hydrogen atoms form hydrogen bonds. Each ..en in a single molecule can form H-bonds with two ..ydrogens (because the oxygen atom has two pairs of nonbonding electrons).

...spondingly become partially positively charged.

Figure shows such a hydrogen bond. The resulting clusters of molecules give water its cohesiveness. In its liquid phase, the network of molecules is irregular, with distorted H-bonds.

When water freezes, the H-bonds form the water molecules into a regular lattice with more room between the molecules than in liquid water; hence, ice is less dense than liquid water.

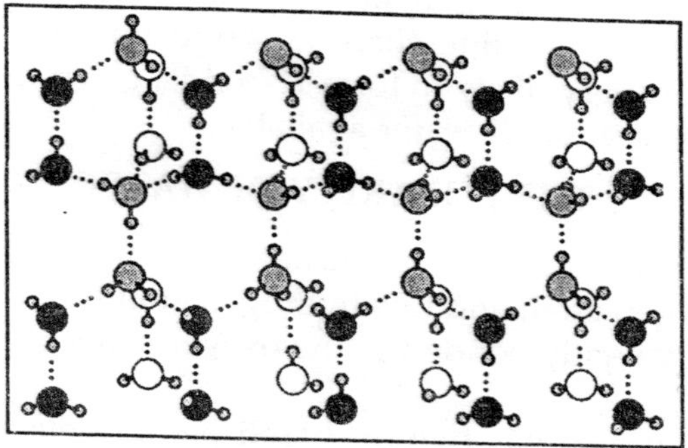

HYDROGEN BONDS AND BIOMOLECULES[H]

In water, the nonbonding electrons are the H-bond acceptors and the hydrogen atoms are the H-bond donors. Biomolecules have H-bond acceptors and donors within them. Consider the side chain of a simple amino acid, serine. The oxygen contains two pairs of nonbonding electrons, as water does, and the hydrogen is correspondingly a focus of partial positive charge. Serine thus can be both an H-bond acceptor and donor, sometimes at the same time.

As you would expect, serine is soluble in water by virtue of its ability to form H-bonds with the solvent around it. Serine on the inside of a protein, away from water, can form H-bonds with other amino acids; for example, it can serve as … donor to the nonbonding electrons on the ring nit… histidine, as shown in Figure.

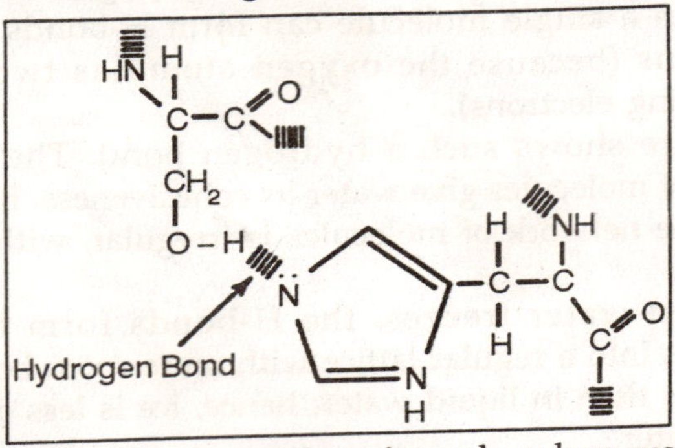

These H-bonds normally exist only when water is not present. If serine's side chain is found on the surface of a protein, it is very likely to form H-bonds, given the relatively high concentration of water available.

THE POLARITY OF WATER

Water has a simple molecular structure. It is composed of one oxygen atom and two hydrogen atoms. Each hydrogen atom is covalently bonded to the oxygen via a shared pair of electrons.

Oxygen also has two unshared pairs of electrons. Thus there are 4 pairs of electrons surrounding the oxygen atom, two pairs involved in covalent bonds with hydrogen, and two unshared pairs on the opposite side of the oxygen atom. Oxygen is an"electronegative" or electron"loving" atom compared with hydrogen.

Water is a"polar" molecule, meaning that there is an uneven distribution of electron density. Water has a partial negative charge ($\delta-$) near the oxygen atom due the unshared pairs of electrons, and partial positive charges ($\delta+$) near the hydrogen atoms.

An electrostatic attraction between the partial positive charge near the hydrogen atoms and the partial negative charge near the oxygen results in the formation of a hydrogen bond as shown in the illustration.

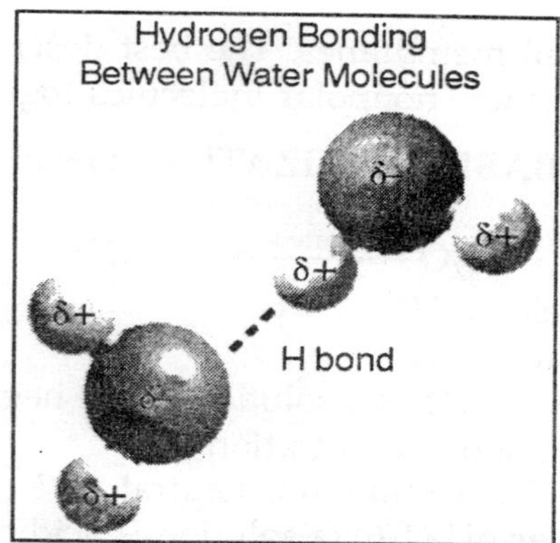

The ability of ions and other molecules to dissolve in water is due to polarity. For example, in the illustration below sodium chloride is shown in its crystalline form and dissolved in water.

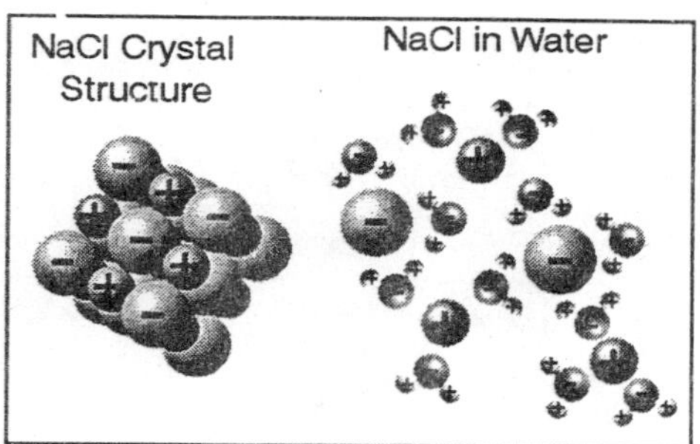

Many other unique properties of water are due to the hydrogen bonds. For example, ice floats because hydrogen bonds hold water molecules further apart in a solid than in a liquid, where there is one less hydrogen bond per molecule.

The unique physical properties, including a high heat of Vapourization, strong surface tension, high specific heat, and nearly universal solvent properties of water are also due to hydrogen bonding.

The hydrophobic effect, or the exclusion of compounds containing carbon and hydrogen (nonpolar compounds) is another unique property of water caused by the hydrogen bonds.

The hydrophobic effect is particularly important in the

formation of cell membranes. The best description is to say that water"squeezes" nonpolar molecules together.

ACIDS AND BASES, IONIZATION OF WATER

$$H_2O \xrightarrow{\text{Limited}} H^+ + OH^-$$

- Acid release H^+
- Bases accept H^+

We define the pH of a solution as the negative logarithm of the hydrogen ion concentration.

- At pH 7.0, a solution is neutral.
- At lower pH (1-6), a solution is acidic.
- At higher pH (8-14), a solution is basic.

Chapter 4

Vitamins

WATER AS SOLUBLE VITAMINS

This manuscript describes the lecture on vitamins contained in the core course entitled,"The Biochemical Basis of Drugs and Diseases. In the first year, our curriculum is designed to focus on systems and diseases. As such, this course acts in concert with both the Anatomy and Physiology course and the Pathophysiology course to present an integrated view of human diseases and systems. The second year of our curriculum is focused on learning about drugs classes and the mechanism of action.

The lecture on vitamins is the last topic of a course focused primarily on metabolism. Because many vitamins play major roles in metabolic cycles, this lecture allows for a brief review of much of the material covered throughout the course. Therefore, vitamin examples are primarily chosen to reinforce major metabolic cycles and also their role in human disease. There is more clinical information that is relevant to vitamins than what is presented in this lecture. However, since first year students in our curriculum have almost no knowledge of therapeutics, the focus of the lecture is on diseases and not clinical practice.

INTRODUCTION

Vitamins are a multibillion dollar industry.1 They are readily available to the public and are the focus of the most frequently asked questions to pharmacists.2 Surprisingly, most pharmacy students receive little training on the many roles of vitamins in nutrition.2 Therefore, this lecture attempts to not only reinforce fundamental biochemical and metabolic pathways in the human body, but also to provide pharmacy students with practical information.

The diet is a vast source of important nutrients. These nutrients include several important classes of biomolecules such as:

- Energy yielding components (carbohydrates, lipids, and proteins).
- Essential and nonessential amino acids.
- Essential fatty acids.
- Minerals.
- Vitamins.

Vitamins are organic substances that must be provided by the diet either because they cannot be biosynthesized or the amount that is provided through biosynthesis is inadequate for maintaining normal health. Vitamins are broadly divided into 2 classes based upon their hydrophobicity. The more hydrophilic vitamins are termed the water-soluble vitamins and are composed of the B-complex vitamins and vitamin C. The more hydrophobic vitamins, referred to as the fat-soluble vitamins, are composed of Vitamins A, D, E, and K. This article focuses primarily on the watersoluble vitamins due to their greater role in the major metabolic cycles, which are the primary focus of this course.

The normal North American diet is sufficient to prevent significant vitamin deficiencies and the related diseases associated with these deficiencies. However, there is increasing concern that slight vitamin deficiencies in a number of water-soluble vitamins (B1, B6, B12, folate and vitamin C) are risk factors for diseases such as such as depression, cancer, cardiovascular disease, and neural tube defects. The prevalence of this slight vitamin deficiency is likely related to diet, since most of these vitamins are supplied by fruit and vegetable intake. A recent survey has shown that only 25% of the population meets their daily intake of 5 servings.3 Food preparation is another source of vitamin loss. For example, heating food for more than 2 hours causes more than a 10% loss of most water-soluble vitamins. Refrigeration, freezing, and reheating have all been shown to lead to further loss of vitamins. Exposure to light causes significant loss of riboflavin from foods and the combination of heating and light can almost completely remove this vitamin from food.

Thiamin (Vitamin B1)

Thiamin functions in the body as thiamin pyrophosphate (TPP) an important enzyme co-factor.5 Thiamin reacts with adenosine triphosphate (ATP) to form thiamin pyrophosphate thiough a reaction mediated by the enzyme thiamin diphosphokinase. Following its production, TPP is incorporated into 2 important enzymes: pyruvate dehydrogenase and ketoglutarate dehydrogenase. Pyruvate dehydrogenase is part of a multi-enzyme complex that acts to convert pyruvate generated in glycolysis into acetyl-CoA for entry into the tricarboxylic acid (TCA) cycle. Thiamin pyrophosphate is also used as a co-factor for the enzyme -ketoglutarate dehydrogenase, which is a key point of regulation in the TCA cycle. -Ketoglutarate dehydrogenase is involved in the conversion of ketoglutarate to succinyl CoA. In the case of both enzymes, TPP assists in decarboxylation of a small ketoacid.

In addition to its uses in metabolism, thiamin may enhance circulation and blood formation. It is required for maintenance of the nervous system and is used in the biosynthesis of the neurotransmitters acetylcholinc and *hydroxybutyrate (GABA). Thiamin also is used in the production of hydrochloric acid in the stomach and, therefore, has a role in digestion.

Thiamin is provided in the diet through most grains. A deficiency of thiamin is called beriberi and it is most often observed in Southeast Asia.6 The main staple of diet in this part of the world is rice, which does not provide dietary thiamin to as great an extent as other grains. The symptoms of beriberi include limb pain, muscle weakness, and low cardiac output. all symptoms are related to the diminished capacity of the major energy producing pathways that are dependent on TPP as a co-factor.

Mild deficiencies of thiamin are sometimes observed in the elderly and in low-income groups on restricted diets. The earliest symptoms of thiamin deficiency are loss of appetite, constipation, and nausea. Other symptoms such as mental depression, peripheral neuropathy, irritability, and fatigue are related to the role of thiamin in maintaining a healthy nervous system.

The most common form of thiamin deficiency in the United States is alcoholic neuritis. Because alcoholics normally have a poor appetite, overall food consumption is low. In addition, alcoholics are predisposed to developing nutritional deficiencies since alcohol is their primary calorie source.

Under the broad umbrella of alcoholic neuritis come two disorders, Wernicke's Syndrome and Korsakoff s Psychosis.7 The symptoms of Wernicke's Syndrome (ophthalmoplegia, nystagmus, and ataxia) respond quickly to administration of thiamin, whereas the more severe memory and learning disorders associated with Korsakoff s Psychosis are refractory to thiamin treatment.

Riboflavin (Vitamin B2)

The reduced forms of these enzyme co-factors can donate these electrons to return to their previous fully oxidized forms. It is this ability to act as a conduit for electron transfer reactions that makes FAD and FMN such important enzyme cofactors.

Riboflavin is available from a wide variety of dietary sources such as milk, cheese, meat, eggs, and cereal products. Symptoms associated with riboflavin deficiency include sore throat, dermatitis, anemia, neuropathy, and cataract formation. Riboflavin deficiency is not normally observed in the United States; however, deficiency may be observed as part of a general case of malnutrition or in cases of chronic alcoholism.8 The conversion of riboflavin to FMN is required for absorption and transport into many tissues.

The conversion has been shown to be inhibited by hypothyroidism and the structurally similar medications chlorpromazine, adriamycin, quinacrine, tetracycline, and tricyclic antidepressants.9-12

Niacin (Vitamin B3)

Niacin can be found in foods such as meats, breads, and beans. Mild niacin deficiencies have similar symptoms to those observed with riboflavin, which is not surprising due to the similar roles both of these vitamins play in biochemical reactions. Niacin deficiencies are occasionally observed in alcoholics, cases of general malnutrition, and in the elderly on restricted diets.

A severe deficiency of niacin is known as pellagra, which is derived from the Italian phrase meaning rough skin. Pellagra is marked by dermatitis and also is notable for causing a blackening of the tongue. Early cases of pellagra were first observed in Europe shortly after the introduction of corn from the voyages of Christopher Columbus. In these cases, poor farmers who were raising corn as animal feed were particularly susceptible. Because niacin in corn is not bioavailable unless treated with a strong base such as lye, the farmers developed niacin deficiencies.

Europeans did not know this processing method until revealed to them by Native-Americans during the early colonial period of North America. However, the connection between pellagra and niacin was not known until the early 20th century. Pellagra was a significant health issue in the United States over the period from 1900-1940 resulting in over 100,000 deaths. Today most diets are supplemented with niacin through enriched flour, which receives its name because of the added niacm.13

Niacin can be administered in doses of 2g to 4g to causes a decrease in circulating levels of cholesterol and LDL. While the cholesterol lowering effects of niacin are desirable there are potential side effects from such large doses of this vitamin. The most immediate reaction observed from large doses of niacin is vasodilation resulting in flushing. Over time there may be a reduction in fatty acid mobilization causing a depletion of glycogen and lipid stores in muscle tissue. Long-term exposure may also elevate blood glucose and uric acid levels, suggesting increased risk for patients who are on the borderline for diseases such as diabetes and gout. Prolonged use of high doses of niacin can lead to elevated levels of the serum enzymes alanine aminotransferase and aspartate aminotransferase, which may suggest liver damage.

Pyridoxine (Vitamin B6)

Pyridoxine is the precursor to the active enzyme cofactor pyridoxal phosphate (PLP). Pyridoxal phosphate is a critical co-factor for enzymes involved in reactions involving many amino acids. The N-terminus of the amino acid forms a covalcnt bond to PLP, allowing a wide variety of displacement reactions to occur at the alpha carbon. These include decarboxylations, transaminations, and transfers of side chains. Pyridoxine, therefore, plays a central role in the production of many neurotransmitters, such as serotonin, norepinephrine, and histamine. Pyridoxine also is important in the production of heme.

The enzyme LAAAD is present in both the gut and in the brain. This presents a problem because L-dopa is readily transported across the blood brain barrier while dopamine is not. Therefore, orally administered L-dopa is rendered ineffective in the gut by LAAAD. Co-administration of carbidopa resolves this dilemma by inactivating LAAAD in the gut. Like dopamine, carbidopa cannot cross the blood brain barrier leaving LAAAD in the brain free to carry out the production of dopamine at the site of action.

In addition to hydrazine containing drugs, penicillamine, used in the treatment of Wilson's disease, cystinuria, and rheumatoid arthritis reacts with and inactivates pyridoxal phosphate.15 Patients treated with penicillamine occasionally develop convulsions, which can be prevented by supplementation with vitamin B6.

Pyridoxine is found in foods such as meats, breads, eggs, soybeans, and many vegetables. A deficiency in pyridoxine can cause facial lesions, depression, peripheral neuropathy, and glossitis.

The neurological complications can be directly linked to the effects on neurotransmitter production. Mild pyridoxine deficiencies are sometimes observed in young women taking oral contraceptives. Deficiencies have also been observed in patients with gasteroenteritis or Crohn's Disease, presumably due to poor absorption. Mild deficiencies are of concern due to a correlation with an increased incidence of breast cancer and to the risk of coronary heart disease.16,17 Neurotoxicity has been noted with doses of vitamin B6 in excess of 500 mg/day.

Cobalamin (Vitamin B12)

Cobalamin is vital for cell growth and replication. Its major site of action is at the interface between the folic acid cycle and the active methyl cycle, where cobalamin is a co-factor for the enzyme homocysteine methyltransferase. This enzyme catalyzes the transfer of a methyl group from tetrahydrofolate to homocysteine, forming the amino acid methionine. These 2 cycles impact many other pathways due to the large number of memylation reactions in the body, especially nucleic acid biosynthesis and neurotransmitter biosynthesis.

Cobalamin is found in meats and dairy products. Deficiencies are not common because the liver can store a 6-year supply of cobalamin. In addition, cobalamin is highly conserved by enterohepatic recirculation. Strict vegetarians (vegans) may take up to 20 to 30 years to develop a deficiency, whereas inadequate absorption from the ileum due to ileitis or loss of a glycoprotein that complexes with cobalamin prior to absorption may take from 2 to 10 years to become symptomatic.

The anesthetic nitrous oxide inactivates cobalamin and can cause patients with marginal serum levels to develop deficiencies within a week.19 While rare, deficiencies in vitamin B12 are severe and manifest in erythrocytes and nervous tissue. In erythrocytes, nucleic acid biosynthesis is slowed. This results in stem cells that are committed to development into erythrocytes possessing twice their normal protein content but lacking the necessary nuclear material for proper cell division. The resulting anemia, termed megaloblastic anemia, can be quite severe.

Cobalamin deficiency can also cause permanent damage to the nervous system. The results are swelling of neurons and demylination of nerve cells, followed by cell death. These progressive manifestations cause a wide range of neurological symptoms including unsteadiness, decreased reflexes, paresthesias of the extremities, and ultimately confusion, memory loss, hallucinations, and psychosis.

The neurological symptoms of cobalamin dcficiency may be mistaken for multiple sclerosis. In the elderly and in alcoholics, cobalamin deficiency should be considered as a possible cause of dementia even in the absence of anemia. The neurological symptoms associated with cobalamin deficiency are not directly related to irregularities in the formation of erythrocytes.20

Nutrient depletion or shortage may arise from inadequate food intake, impaired digestion, inefficient absorption or malabsorption, inefficient assimilation and other physiologic or functional disorders.

Genetic errors may interfere with or prohibit assimilation and utilization. Gastrointestinal motility disorders may have similar effects. Stress may affect both gastrointestinal tract transit times and quantities or ratios of nutrient needs, leading to depleted nutrients. Americans now spend $2.5 billion yearly on vitamins and minerals, up from $328 million 25 years ago. Roughly 55% of women and 45% of men take some kind of vitamin supplement.

Most of the vitamin supplements consumed by Americans, including those used in"enriched" foods, are making them more deficient! The greatest and most enduring of medical discoveries of the twentieth century were the vitamins. Here, at last, was the discovery of factors that caused health. Vitamin A (1912, the first vitamin to be discovered) was named retinol because, without it, a healthy retina in the eye could not be formed. Then it was found to cure night blindness. And that was just the beginning.

Along came the B-vitamins and the cures for beriberi, pellagra, pernicious anemia, nerve degeneration, enlarged heart, energy production, and much more. These came from unprocessed, whole, natural foods. When man Realised his foods had been devitalized from processing and refinement, he decided to replace the lost vitamins. But, in his effort to extricate himself from the refined food trap,"scientific confusion" led him to try to replace them with"chemical, man-made vitamins." So, again, he refined down natural vitamins into pure, crystalline vitamins, thus robbing them of important food values, just as he had refined natural grains and sweets into their pure, crystalline, carbon-white flour and sugar.

Then, to save money, he duplicated these crystalline vitamins synthetically, mostly from COAL-TAR products. Thus, natural vitamins as well as natural foods were robbed of nature's life and valuable elusive nutrients.

The latest round in conventional medicine's ongoing attempts to discredit (and ultimately outlaw) nutritional supplements is found in a highly questionable study published February, 2007 in the Journal of the American Medical Association, which claims that vitamins actually increase the risk of death.

The study claims to have Analysed a collection of previous studies on Vitamin A, beta carotene, Vitamin C, Vitamin E and selenium, concluding that most of the nutrients are actually dangerous to human health. Of course, this is research from conventional medicine - an industry that promotes patented chemicals as perfectly safe, even though FDA-approved pharmaceuticals are killing at least 100,000 Americans each year. (Imagine the uproar if vitamins killed even a fraction of that number)

To avoid getting hoodwinked by questionable research on"vitamins," you have to strongly consider the financial interests of the source of this research. JAMA accepts millions of dollars in advertising from drug companies each year, and its pages are absolutely packed with drug ads. The American Medical Association, for its part, has long worked to discredit alternative medicine and has even been found guilty by U.S. federal courts of engaging in a conspiracy to destroy chiropractic medicine. The AMA, which is largely considered a joke by anyone familiar with natural health, is hardly a credible source for publishing scientific findings on nutrition. To protect the multi-billion dollar drug industry, it's likely the AMA would say practically anything.

This research published in JAMA does remind us of one important point, however: synthetic chemicals are harmful to human health. If you take cheap"vitamins" made of these synthetic chemicals, you are doing yourself more harm than good. These cheap vitamin manufacturers, by the way, are usually owned by pharmaceutical firms.

Conventional medicine researchers try to blur the line between"junk vitamins" and"quality vitamins" by classifying ALL nutritional supplements as"vitamins," regardless of what they're really made from. By discrediting a few synthetic chemicals, they can effectively dissuade the masses from taking ANY vitamins, including the good ones. And that is, of course, their goal: to use F.U.D. (fear, uncertainty and doubt) to scare consumers away from nutritional supplements so that patients will flock to the patented, synthetic chemicals that earn drug companies billions of dollars in profits. Drugs make money for Big Pharma, and vitamins compete with drug sales.

Round up 100 people who are taking multiple pharmaceuticals, and compare their health to 100 people who are taking vitamins and nutritional supplements. Guess who's healthier? The supplement crowd will be healthier every time. It's the obvious question: If vitamins are so dangerous, where are all the dead vitamin takers? And if pharmaceuticals are so safe, where are all the super-healthy prescription drug patients? They are nowhere to be found.

The healthiest people, by far, are those who take supplements, who engage in regular exercise, and who avoid taking prescription drugs.

Only fools believe research about nutrition that comes from the American Medical Association or its journal. Conventional medical researchers declaring that vitamins are worthless is about as credible as Bush Administration climatologists claiming there's no such thing as global warming. With the publication of this research, the distortion of health reality is now complete. According to the Medical

Association, vitamins will kill you but pharmaceuticals will make you healthy.

DESCRIPTION ABOUT VITAMINS

It is a complex of biological"wheels within wheels" consisting of enzymes, coenzymes, antioxidants and trace mineral activators. Enzymes, being protein, contain amino acids and trace minerals. Every mineral and amino acid needed by living cells may be found in a natural assemblage of vitamin concentrates obtained from wholesome foods. Identifying a vitamin in terms of a single chemical structure is self-defeating--you need the whole complex to get the vitamin function.

The natural vitamin complex carries trace mineral activators and without them, the vitamin fails in its metabolic function. Copper in its organic, biologically active form tyrosinase, is the trace element activator of the vitamin C-complex. Likewise, manganese activates the B complex. In turn, selenium is the trace mineral activator of the vitamin E-complex. Enzyme activators, destroyed by heat, steam sterilization and pasteurization, include Manganese, Cobalt, Molybdenum, Zinc, Selenium and Vanadium...and are effective only in the organic state.

By reducing a vitamin complex to a single chemical, chemists can claim to duplicate it in a Labouratory and easily mass produce it and sell it as an additive to food or a pill. This wouldn't be possible if they were charged with keeping all the synergistic components intact. As such, vitamins are given single chemical names and single chemical structures for the purpose of selling synthetic vitamins.

The failure of the pharmaceutical approach to vitamins originated in the chemists' synthetic isolation of aelements, uncombined with its biological matrix. The discovery of the vitamins coincided with the boom in industry, agriculture, and chemistry, which were crowding their way into all facets of life. But in the attempted duplication of living things, like artificial fertilizers or vitamins, chemistry had its most influence. The biological activity of synthetic and natural vitamins is not identical. Anyone who has studied anatomy, physiology, neurology and biochemistry will readily admit that the human body is a neuro-biochemical plant--a composite of tissues, fluids and organs that exist in a living state by virtue of neuroelectric and biochemical reactions.

Being biochemical in nature, the body operates on biochemicals--LIVE CHEMICALS--in contrast to dead or lifeless chemicals. Those biochemical factors are derived from whole, natural, uncooked and unprocessed foods. Cooking and processing alters biochemicals to dead chemicals. The human body cannot properly function if it is fueled with dead chemicals rather than live, complex biochemicals.

The difference between the live biochemical food factor and the dead chemical (counterfeit food) is dependent upon two primary requisites:

- Unaltered proteins.
- Viable enzyme systems.

Dr. Casmir Funk, who discovered the method for concentrating vitamin B, and the one who coined the term, Vitamine, said,"The synthetic product is less effective and more toxic." Every process used to extract vitamins from foods, destroys the vitamin complex. What you end up with is similar to a banana peel without the banana. When you take these incomplete supplements, they rob from the body's nutritional reserve what is missing (the banana), to try to complete the complex.

This creates what is called a rebound deficiency. If you find the shed skin of a snake, you haven't found a snake. In theBiomedical Nitty Gritty, Vol. 3, No. 1, Jan. 1982, Dr. Richard Murray D.C., presented an abundance of incontestable evidence that there is a tremendous difference between natural and synthetic vitamins..."Synthetic vitamins are reverses conformationally or mirror images of natural vitamins. However, synthetic vitamins are not vitamins at all but synthesized fractions of a vitamin complex...a mirror image duplicate of just a portion of the real, biologically active and physiologically-precise, nutritional complex. There is no way that a fraction of a vitamin can be called a vitamin." He continues,"Textbooks record that vitamin B1 was first isolated in the year 1926. About 10 years later, scientists were able to synthesize the mirror image of Thiamine from coal-tar.

About the beginning of WWII (Sept.-39) enrichment of refined, devitalized flour became compulsory. Bleached wheat flour has more than 30 known nutrients removed with 4 synthetics added back in. Today, the majority of people in the civilized world are becoming victims of degenerative disease resulting from chronic poisoning and malnutrition.

Even though well fed, they are starving to death for the most essential micronutrients. Since World War II, the American people and people of other countries as well, have had a daily ration of a genetic poison in most of the bread, flour products, cereals and other food items that are forced, by law, to be enriched with synthetic vitamins.

Then, after there develops a suspicion that these'foods' are inadequate to support life, modern advertising science steps in to propagandize the people into believing that there is nothing wrong with counterfeit or devitalized foods, that they are products of value...and...the confused public is totally unable to arrive at any conclusion of fact. In turn, they continue to buy the rubbish that is killing them off, years prematurely."

Gilbert Levin discuss optical isomers or asymmetrical molecules with atomic arrangements that are mirror images of each other, states:"Because its structure is reversed, a left-handed molecule can't take part in chemical reactions meant for a right-handed molecule any more than a left hand can fit in a right-handed glove--its odd geometry prevents it from being metabolized by the body." In turn, the human physiology cannot properly utilize synthetic (mirror-image) fractions in the way natural complexes find their way into the biological reactions that are essential to tissue repair and the sustenance of life. A synthetic vitamin fraction can only be utilized for a drug or pharmacologic effect. Drugs are poisons, by definition. All pharmacologic therapies induce side effects and high-dose nutritional therapies are no exception. Only synthetic supplements achieve drug status. The effect of a drug is palliative not curative. Palliation means a masking or covering-over of symptoms, while the disease process remains unchanged or progressively gets worse for lack of attention."It is better to have no vitamins at all than a mixture of crystalline-pure vitamin fractions, with one or more members of the complex missing."

Natural complexes differ from synthetic or crystalline vitamins in many ways:

- Natural complexes are colloidal, protein in nature and usually exist as enzymes or coenzymes.
- The crystalline vitamin can't be separated from its protein part without destroying its biological activity.
- The natural"live" complex carries trace mineral activators...without which the enzyme fails as a biochemical catalyst.
- Crystalline vitamins, taken into the body, must recombine (if possible) with their natural required components (robbed from the body's reserve) before they can function as a vitamin. Most, if not all of the crystalline component is lost through the kidneys.

Vitamins are not meant to be eliminated in urine any more than silverware is to be thrown out with the leftovers. Since most poisons stimulate circulation in the human body for the purpose of moving blood through the kidneys, more often to filter out the said toxin as quickly a possible...the speeded-up circulation also delivers more blood and more oxygen to the brain with a created sense of euphoria. Natural and low-potency vitamin complexes do not and will not produce euphoria, any more than eating a (non-poison-sprayed) carrot.

A synthetically-derived substance may cause a reaction in a chemically susceptible person when the same material of natural origin is tolerated, despite the two substances having identical chemical structures. The layman who doesn't understand the biochemical nature of true nutritional changes, is easily deceived by the pseudo-nutritional therapist and believes the initial response (drug effect), to the massive dose of one or more prescribed megavitamins, represents that what is being taken is exactly what is needed.

Nutritional therapists know after a time on synthetic vitamin fractions, the biochemical reaction is reversed, the euphoria ends and the original symptoms return; except amplified by vitamin poisoning. Also, the diagnosis and evaluation of symptoms are now clouded by the weeks, months or years of improper supplementation. Why the great mystery about all of this?

Why don't doctors know about vitamins or nutrition in general? One big answer was revealed in the Journal of the Medical Association. It honestly admitted:"Nutrition has been neglected by the medical profession." Most medical schools devote less than three hours of total instruction in nutritional deficiency and therapy.

In short, Physicians in the United States are not required to understand nutrition to be licensed to practice medicine. It's not hard to understand why, when medical experiments are performed with synthetic vitamin fractions, these counterfeits fail to produce impressive results. But a reductionist seeks to understand the whole by dismantling its parts. This precludes the wholistic principle of synergism, that the whole is greater than the sum of its parts. For example, what if we tried to understand the properties of water (H2O) by testing hydrogen separately and oxygen separately. Presumably, we would learn enough about each to understand what their combined characteristics would be. Combined as 2 parts hydrogen and 1 part oxygen, H2O will extinguish fire. Separately, the elements are among the most flammable and explosive elements in the universe.

The functions are exactly the reverse in their isolated state from their organically combined state. Similarly,vitamins function as biological mechanisms only when whole and complete, combined with their synergists, as in whole food. Isolated into synthetic chemicals, they fail as catalysts.

CATEGORIES OF SUPPLEMENTS

The following is a summarization of the basic differences between the four categories of supplements available today.

Pure, Hypoallergenic or Natural

This is the most prevalent category of professional

supplements available today. Contrary to popular belief, the vitamins and minerals used to create supplements in this category are not derived from consumable food or botanical sources.

Instead, they are synthetically produced, fractionated or isolated from refined raw materials such as crushed rock, petroleum by-products and organic solvents to produce a pure, crystalline vitamin or mineral analogue. Although the final compound is an analogous structure to a specific vitamin or mineral, it is no longer attached or associated with any of the synergistic co-factors that the nutrient is inherently found within foods. As a result, pure vitamins and minerals have significantly reduced biological activity and a lower rate of utilization in the body.

Food-Based

This category of supplements is created from a physical mix of pure, fractionated nutrients with varying amounts of food powders or extracts. There are several different factors that need to be considered when determining the absolute nutritional value of a food-based formula. The primary factors that affect the quality of a food-based supplement are the amount, variance and form of the foods included in a particular formula.

A premium food-base formula is one that provides a broad range of unadulterated whole food or plant concentrates (preferred over powders) in at least a 3:1 ratio of food to nutrients in a formula will significantly improve the bioavailability and biological activity of the vitamins and minerals.

Delivering isolated vitamins and minerals in a premium food base is an energy-saving process, as the body will not waste vital energy stores to transform the vitamins and minerals into absorbable nutrients. Ascorbic acid is a good example of a chemically synthesized vitamin that has been vastly studied; it has demonstrated enhanced utilization when the nutrient is delivered with its natural co-factors, such as bioflavonoids, rutin and tyrosinase, which always co-exist with vitamin C in food. The utilization factor can be increased by as much as 3 to 5 times (from 10% for vitamins in the first category, to 3 – 50% for food-based).

To determine if a supplement has a food-base, review the label for chemical names, as well as a combination of food extracts or powders. To determine the ratio of nutrients to foods, add up the total milligrams of the listed nutrients and for the food extracts.

Biocultured Supplements

Biocultured supplements are a physical mix of food, pure isolated nutrients and may have one or more strains of yeast and/or probiotics. Once combined, this mixture is allowed to ferment in a controlled environment.

The process of fermentation, also known as culturing, biologically alters the biochemistry and nutrient content of the food. Cultured foods have been shown to have numerous health benefits and have played a role in the diet of many cultures for centuries; however, the body cannot sustain optimal health on cultured foods alone. This is due to the loss of substances, nutrients or compounds that are protective of health, such as vitamin c. Depending on the strains used during the culturing process, the generation or inactivation of toxic substances may also occur. In addition, many proteins and enzymes are denatured during the culturing of biological activity of the nutrients.

Supplements that fall into this category will denote the foods on which the yeast and/or probiotics feed. Not all cultured supplements list the isolated vitamins and minerals used in a particular formula.

WHOLE-FOOD SUPPLEMENTS

Many companies tout that their supplements are made from whole-foods; however, there are only a few companies that actually can substantiate this claim. Authentic whole food supplements are made with vitamins and minerals delivered in whole-food concentrates. The food concentrates selected have been grown in a controlled environment where budding plants are fed a nourishing broth enriched with bioactive peptide carriers. The peptide carriers act as chaperones, which govern a vitamin's or mineral's capacity to store the nutrient within its cells is successfully achieved, the growing process is complete, and harvesting is commenced using proteolytic enzymes to break down the cells' walls, and release the intra-cellular material.

The whole-food extract or concentrate produced is rich in a particular vitamin or mineral and is delivered in a synergistic state, with all of the plant's inherent food attachments intact and unadulterated. These attachments are vital co-factors facilitating the body's natural ability to recognize and utilize the nutrients present within the concentrate. Whole-food nutrients are recognized by the body as food, and therefore are better utilized and retained by the body when compared to any other supplemental form. Food is the complex form in which all nutrients are found in nature.

Supplements in this category will list the nutrient names--vitamin C, folate, calcium, etc.--and the mg. dosage of each, butthey will not denote a fractionated chemical name. Nutrients as they exist in food cannot be listed as one identifiable organic compound in this way, for the reason that they always exist within an interconnected weave of parts, and never as one isolated element. Nutrients in a whole-food state will be listed with their food source in the supplement facts panel such as vitamin C.

Bioavailability Spectrum

Let's take a closer look at the bioavailability of supplements on a spectrum, from least bioavailable to mostbioavailable. On the low end of the spectrum you would have an inexpensive, store bought, hard-packed tablet. Something brightly Coloured that you can't break with your fingers. Such supplements have many fillers and binders, including ingredients that can be found in wood putty and tar. You can't take these on an empty stomach, or else you'll get an upset stomach. These are difficult to digest and breakdown, and often times your body doesn't at all. And these have been known to have been passed whole! For that reason you can understand why they have a bioavailability of 0-25%. So, just as a reminder, when you look at the daily values of nutrients on the label of these products, your body is not getting those Percentages of nutrition, but rather up to 25% tops.

Moving up the bioavailability spectrum we come to the capsule delivery system. They have no fillers or binders, and you usually have to take larger quantities of these because they are not very concentrated. The bioavailability is a little better, but still only 40-60%.

Getting higher up the bioavailability spectrum brings us to liquid supplements. These are better then the previous two in bioavailability and range between 80-90%, but many liquid herbs and oils do not taste very good.

Finally we arrive to isotonic-capable supplements. Isotonic supplements are the most bioavailable delivery of supplements and are up to 95% absorbable.

Isotonic-Capable Supplements

It is the proper ratio of nutrients to water so that it can be seamlessly integrated into the blood stream. An isotonic capable supplement works similarly. It is the proper measurement of nutrients to water to create an isotonic capable solution. When it is taken on an empty stomach, it is recognized as the same pressure as the stomach. The pyloric valve is signaled to open, and the full concentration of nutrients is delivered to the small intestine (our absorption site) where it is up to 95% absorbable within 5-15 minutes of consumption.

It is important to note that not all liquids are isotonic capable, again it is a specific ratio of medicine or nutrients to water which makes it isotonic. Everything that you ingest and put into your stomach has to be digested down to an isotonic state before it is released to the small intestine. When you eat some food and take a standard pill or capsule vitamin, it all remains in your stomach digesting for anywhere from 40 minutes to 4 hours. At that point a certain amount of dilution of nutrients has taken place. Some low quality pills do not even break down due to harsh fillers and binders, but still pass to the small intestine where none of the content of the supplement can be absorbed. This is why the bioavailability of pills and capsules are so significantly lower than isotonic capable supplements.

Vitamin-A

Vitamin-A was the first fat-soluble vitamin to be discovered. In 1919, it was recognized that egg yolk contained a fat-soluble substance essential for life. In 1924, it was demonstrated that the eye disease, xerophthalmia, in children could be prevented by feeding them butter or cod liver oil. Vitamin-A was purified--separated from its natural constituents--and its refined chemical structure proposed in 1931. Synthesis, an artificial chemically-pure duplication, was accomplished in 1947 at Hoffman-LaRoche, a pharmaceutical company in Switzerland. Vitamin-A is actually a whole family of compounds, including retinol, retinal, retinoids, carotenes, and carotenoids, and always occurs in nature with synergists such as fatty acids, chlorophyll, other vitamins, enzymes, minerals, and trace elements. Retinol, retinal, and retinoids are found in foods of animal origin such as eggs, liver, fish, butter, and cheese.

The scientific term for vitamin-A is retinol, because of its presence in the retina of the eye. The role of retinol in vision was elucidated by a number of brilliant scientists, beginning in 1877 with a German, W. Kuhne, who discovered that the purple retinas from dark-adapted frogs turned yellow when exposed to light. The purple Colour is restored in a complex biochemical cycle involving vitamin-A, which makes vision possible.

Other scientists demonstrated the role of vitamin A in cell differentiation, bone development, reproduction and immune system function. Dr. Weston Price confirmed the value of vitamin-A in traditional diets during his studies of primitive peoples carried out during the 1930s and 1940s. Due to the outstanding scientific work of these and many other researchers, the administration of cod liver oil to growing children--a tradition found among Arctic peoples such as the Scandinavians and Eskimos--became standard practice until after the Second World War.

In recent decades, much vitamin-A research has focused on its role in preventing cancer, and its use in combination with nontoxic therapies in the treatment of cancer. Dr. Max Gerson treated many cases of terminal cancer with excellent results, using raw liver juice, a rich source of vitamin-A. Modern confinement farming practices effectively prevent vitamin-A from incorporation into animal foods and the processing industry would rather use vegetable oils than animal fats. Some vegetable oils contain carotenes but they do not contain true vitamin-A. Only animal fats contain vitamin-A and vitamin-A is present in large amounts only when the animals have a source of carotenes or vitamin-A in the diet, such as green pasture, insects, and fish meal.

Many popular books on nutrition insist that humans can obtain vitamin-A from fruits and vegetables. Even worse, FDA regulations allow food processors to label carotenes as vitamin-A. Vitamin-A from butter may be two or three times as potent, unit for unit, as vitamin-A from cod liver oil in some people.

The food industry, and the low-fat school of nutrition that the industry has spawned, benefit greatly from the fact that the public has only vague notions about vitamin-A. In fact, most of the foods that provide large amounts of vitamin-A--butter, egg yolks, liver, organ meats and shellfish--have been demonized. Under optimal conditions, humans can indeed convert carotenes to vitamin-A. This occurs in the upper intestinal tract by the action of bile salts and fat-splitting enzymes.

Of the entire family of carotenes, beta-carotene is most easily converted to vitamin-A. The accepted ratio of how many units of beta-carotene is needed to produce one unit of vitamin-A is 6:1 and some research suggests an even higher ratio. This means that you have to eat an awful lot of vegetables and fruits to obtain even the daily minimal conversion. But the transformation of carotene to retinol is rarely optimal.

Strenuous physical exercise, periods of physical growth, pregnancy, lactation and infection are stresses that quickly deplete vitamin-A stores.

Diabetics and those with poor thyroid function, which could include at least half the adult U.S. population, cannot make the conversion. In fact, the thyroid gland requires more vitamin-A than the other glands, and cannot function without it (mercury toxicity diminishes the thyroid and pituitary glands' functions). And a diet rich in vitamin-A will help protect the diabetic from the degenerative conditions associated with the disease, such as problems with the retina and with healing (mercury binds with the sulfur in insulin).

Children make the conversion very poorly and infants not at all--they must obtain their precious stores of vitamin-A from animal fats--yet the low-fat diet is often recommended for children. Vitamin-A from animal sources (retinol) is primarily absorbed within three to five hours after ingestion in the upper intestinal tract; it is here that the fat-splitting enzymes and bile salts convert carotenes into usable nutrients, though conversion and assimilation takes up to six or seven hours. Carotenes are converted by the action of bile salts (mercury inhibits production of taurocholic acid, a bile salt) and very little bile reaches the intestine when a meal is low in fat.

The conversion of carotenes is stimulated by thyroxine, a hormone secreted by the thyroid gland. Zinc, iron, and vitamin-E status also influences carotene conversion. Once converted, the carotenes are absorbed in the same way as retinol and allocated by the liver. Unabsorbed carotene is excreted in the feces.

Butterfat stimulates the secretion of bile needed to convert carotenes from vegetables into vitamin-A, and at the same time supplies very easily absorbed true vitamin-A. Polyunsaturated oils also stimulate the secretion of bile salts but can cause rapid destruction of carotene unless antioxidants are present. It is very unwise, therefore, to depend on plant sources for vitamin-A. This vital nutrient is needed for the growth and repair of body tissues; it helps protect mucous membranes of the mouth, nose, throat and lungs; it prompts the secretion of gastric juices necessary for proper digestion of protein; it helps to build strong bones and teeth and rich blood; it is essential for good eyesight; it aids in the production of RNA; and contributes to the health of the immune system.

Strenuous physical exercise, excessive consumption of alcohol, excessive consumption of iron (especially from"fortified" white flour and breakfast cereal), use of a number of popular drugs, excessive consumption of polyunsaturated fatty acids, zinc deficiency and even cold weather can hinder the conversion of carotenes to vitamin-A, as does the low-fat diet.

Natural vitamin-A provided by liver, eggs, butter, cream and cod liver oil is well recognized as providing excellent protection against birth defects. Vitamin-A deficiency in pregnant mothers results in offspring with eye defects, hydrocephalus, displaced kidneys, harelip, cleft palate and major malformations of the heart and large blood vessels. Vitamin-A stores are rapidly depleted during exercise, fever and periods of stress. Vitamin-A deficiencies are widespread and contribute to high infant mortality, blindness, stunting, bone deformities and susceptibility to infection.

These occur even in communities that have access to plentiful carotenes in vegetables and fruits. It is required for cellular differentiation (determines the function that cells will have): this assures that the cells which are lost from natural turnover, stress, insult, injury, disease, etc. are reproduced in the exact same form as the ones being replaced. Growing children actually benefit from a diet that contains considerably more calories as fat than as protein. Generous amounts of vitamin-A insure healthy reproduction and offspring with attractive wide faces, straight teeth and strong sturdy bodies.

This vitamin assists in normal pregnancy, embryonic development, successful reproduction, fertility, lactation (nursing), and reproductive organ function, spermatogenesis, to adrenal, thyroid, and other gland functions. Vitamin-A is important for normal eyesight; necessary for night vision. A high-fat diet that is rich in vitamin-A will result in steady, even growth, a sturdy physique and high immunity to illness.

Vitamin-A-rich foods like liver, egg yolk, cream and shellfish confer resistance to infectious diseases in children and prevent cancer in adults.

Children with measles rapidly use up vitamin-A, which can result in irreversible blindness. An interval of three years between pregnancies, allows mothers to rebuild vitamin-A stores so that subsequent children will not suffer diminished vitality. Kwashiorkor is as much a disease of vitamin-A deficiency, leading to impaired protein absorption, as it is a result of absence of protein in the diet. High-protein, low-fat diets are especially dangerous because protein consumption rapidly depletes vitamin-A stores. Children brought up on high-protein, low-fat diets often experience rapid growth.

The results--tall, myopic, lanky individuals with crowded teeth, and poor bone structure, are commonplace in America. High-protein, low-fat diets can even cause blindness. Scarcity of good quality raw dairy products, a rejection of organ meats as old fashioned or unhealthful, and a substitution of vegetable oil for animal fat in cooking all contribute to the physical degeneration and suffering of Third World peoples. Supplies of vitamin-A are so vital to the human organism that mankind is able to store large quantities of it in the liver and other organs. Even people who can efficiently convert carotenes to vitamin-A cannot quickly and adequately replenish vitamin-A stores from plant foods.

Many factors interfere with its absorption and utilization. Inadequate fat in the diet, poor production of bile salts, low enzyme status, and compromised liver function can all interfere with the uptake and usage of vitamin-A, especially when given as a supplement in the form of retinol, rather than as a component of whole foods. One tablespoon of cod liver oil contains at least 15,000 IU and one serving of liver (should be from organically raised animals) can contain up to 40,000 IU vitamin-A.

Foods high in vitamin-A are especially important for diabetics and those suffering from thyroid conditions. Weston Price considered the fat-soluble vitamins, especially vitamin-A, to be the catalysts on which all other biological processes depend. Efficient mineral uptake and utilization of water-soluble vitamins require sufficient vitamin-A in the diet.

Synthetic Vitamin-A

Synthetic vitamin-A interferes with the proper utilization of natural vitamin-A from foods. Pure retinol is added to many fabricated foods like margarine, white flour, extruded breakfast cereals and pizza. A synthetic retinoic acid derivative,isotretinoin (Accutane) is used to treat acne as well as psoriasis, cancer, and oral leukoplakia. Children whose mothers took Accutane during pregnancy were born with nervous system defects, brain damage, congenital heart disease, and more. Depression, crying spells, malaise, and forgetfulness are reported human symptoms.

Inflammation of the lips (cheilitis) develops in 90% of the users; conjunctivitis develops in 40%. Muscle damage has also been added to the list. The synovial membranes in joints are supported by vitamin-A.

The vitamin-A complex is supportive to the adrenal glands which produce cortisone and other needed hormones; and vitamin-A plays an important role in any inflammation and repair process. Vitamin-A is important to the external and internal mucous membrane tissues as well as to the immune system. However, synthetic retinoic acid metabolites do not reveal any specific supportive mechanism and results are not seen with synthetic, altered fractions.

The whole vitamin-A complex is known to be very functional, but when distilled or synthetic fractions are used, roles and results are questionable at best. Megadoses of synthetic vitamin-A can result in hypertrophy (excessive growth) of the liver and spleen, with elevation of AST (aspartate aminotransferase), a liver enzyme. Liver fibrosis (abnormal formation of fibrous tissue), fat deposition, obstruction of portal blood flow with portal hypertension (high blood pressure) and sclerosis(hardening) of the blood vessels have been found at liver biopsies.

Excess synthetic vitamin-A is also associated with calcium deposits in soft tissues, contributes to attacks of gouty arthritis, hair loss, irregular menses, emotional lability, insomnia, restlessness, night sweats, and more. Vitamin-A is essential to spermatogenesis, the production of sperm in men, and it requires alcohol dehydrogenase, an enzyme to convert retinol to retinal.

This enzyme may require the trace element, zinc, as well. Excessive doses of synthetic fractions of vitamin-A can convert to a compound that can be more damaging than a deficiency in its effect on the germinal epithelium of the testes. Vitamin-A is present in almost twice the amount in seminal fluid of highly fertile men compared with that of men deficient in sperm. Zinc increases vitamin-A uptake by the seminiferous tubules of the testes. Vitamin-A is bound to a protein, called retinol-binding protein.

The necessity of enzymes, trace mineral activators, amino acids, and other constituents stresses the point that the whole food complex is important and effective whereas an isolated portion cannot produce the same results. Excessive fractionated"vitamins" A and E may produce testicular hemorrhaging. Zinc is required to maintain normal concentrations of vitamin-A in the blood plasma.

Diabetic persons tend to have decreased levels of zinc. Since a zinc deficiency may lead to difficulties mobilizing vitamin-A from the liver for use elsewhere in the body, vitamin-A deficiency will worsen the condition. Another essential constituent is fat, as vitamin-A occurs in foods with fatty acids and is a fat-soluble vitamin. It is absorbed much better when natural, unaltered fats are also present. Other fat-soluble vitamins occur with and work with whole vitamin-A complex such as vitamin E complex and essential fatty acids.

When vitamin E is deficient, one develops low blood levels of vitamin-A; when vitamin E is restored to the diet, vitamin-A levels become normal. Yet, using chemically-pure fractions of"vitamins" A and E in large doses creates more problems, more imbalances. Fractionated alpha-tocopherol (so-called vitamin-E) can supplant vitamin A in the body, reducing its effectiveness. Administration of the whole food complex brings excellent responses.

Stubborn cases of gastrointestinal irritation and even ulceration in the digestive tract that failed to respond to other treatments have totally resolved after use of real vitamin-A complex. Persons with cystitis (inflammation of the bladder) experience relief after a relatively short time taking food concentrates of Vitamin-A and other supportive nutritional complexes. Edema caused by liver pathology is another instance where vitamin-A complex is of particular value; the liver is the chief location of vitamin-A stores. Colds, chronic rhinitis, sinusitis, psoriasis, and many other conditions--especially those involving epithelial tissues--are responsive to the whole vitamin-A complex.

A new strategy that has been the development of and promotion of genetically engineering rice to produce carotenes. Golden rice containing carotenes can't provide true vitamin-A to the world's children but it will further the trend of pushing their parents off the farm and into ghastly slums. The most effective response that consumers have is to boycott processed foods!

Vitamin B-Complex

The B-COMPLEX factors naturally divide themselves into two groups. One group consists of the factors that are alcohol and water soluble. This includes B1, B12, Pantothenic Acid and Adenine, but the primary factor in this group is called Vitamin B4, an anti-paralysis vitamin. It is one that has not yet been recognized by the FDA as being necessary in human nutrition. They say it helps pigeons, but not people. So vitamin B4 is a nerve-promoting factor, it promotes the transmission of nerve impulses. This would be needed with heart conditions involving irregular heartbeat, fibrillation, or heart blocks.

The B4 factor also causes vessels to constrict. It's very useful for people with low blood pressure--the person whose blood vessels are just hanging open, with no tone. The B4 puts strength in the vessels and snaps them into action by improving the transmission of nerve impulses to the vessels. For this reason, vitamin B would be contraindicated in cases of organic heart trouble, such as angina. Here you would need the G factor. The G factors are not soluble in absolute alcohol. Complex G contains the lipotropic factors that metabolize fat.

Complex-G contains B2, B3, B6, Choline, Betaine, Inositol, Folic Acid, Biotin and PABA, and is a vasodilator--it opens up the blood vessels by relaxing the nerves in the vessels. So complex G is for a person with high blood pressurethe tense, coronary type person. It is important to have these factors separated and useable either by themselves or together. This benefit is not achieved with megadoses of synthetic so-called Vitamin-B. Today, the average practitioner of the medical arts, if sufficiently brain washed by the medical-drug-food-dairy-chemical combine of disinformation, cannot conceive that any person could suffer from B-complex deficiency.

Even if they know that refining and processing of food, chemical fertilizers and food additives, all diminish vitamin content and nutrient quality. Because these"foods" are"enriched" with synthetic vitamins. University biochemists and leading health authorities parrot the phrase:"There is no difference between natural and synthetic vitamins." Below are diagrams of the structural formulae of NATURAL vitamin B1 (left) and synthetic B1 (right).

Show any high-school chemistry student with a passing grade the diagrams and tell them they are the same, and that the body does not know the difference. They will laugh in your face. Now, leading nutritionists from the university fountainheads of knowledge will tell you:"there is no difference."

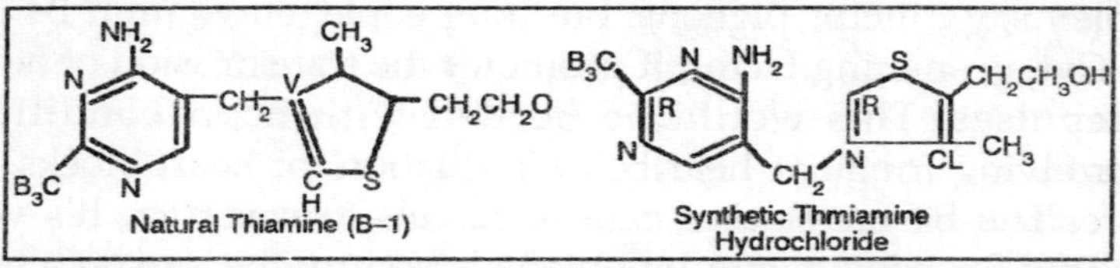

Natural Thiamine (B–1)

Synthetic Thmiamine Hydrochloride

Thiamine is one of two vitamins in the body that contains a sulfur entity; the other is biotin. The high binding affinity that mercury has for the sulfur-containing molecules (thiols and disulfides) gives it the biochemical potential to affect the functions of this critical vitamin and affect critical energy processes in the body.

Not only does thiamine have a sulfur atom in its structure to start with, but its involvement as a coenzyme in critical biochemical processes is connected through transfers involving sulfur molecules.

Out of about two-dozen substances produced by the metabolism of thiamine in the body and excreted in the urine, science has only identified the function of six of them thus far. There is a lot going on metabolically with vitamin B1, that mercury can affect, that haven't even been considered as potential problems because the complete biochemical function of the vitamin has not been identified.

Thiamine is involved in many enzyme reactions in the body. As a coenzyme, B1 is involved in the intermediary metabolism of carbohydrates in all cells of the body. For natural vitamin B1 to perform as a coenzyme, it must be phosphorylated or combined with two molecules of phosphoric acid. As such, it then functions as thiamine pyrophosphate (TPP).

In the biochemically active form, it is also called diphosphothiamine (DPT). Better textbooks of biochemistry use the TPP designation as preferable. Early symptoms of thiamine deficiency, are intriguingly similar to mercury toxicity and are non-specific and hard to pinpoint: fatigue, anorexia, weight loss, gradual loss of muscle strength, peripheral neuritis (numness or increased sensitivity, tingling in the extremities, loss of reflexes), irritability, confusion, depression, lack of initiative, poor memory, gastrointestinal problems (including constipation), and low body temperature.

Cardiovascular symptoms can include edema of the ankles, feet and legs, palpitation, tachycardia, abnormal electrocardiograms, decreased blood pressure, and difficulty in breathing after exertion.

Thiamine participates in the metabolism of tryptophan to form nicotinic acid. Because of this, a deficiency of thiamine can cause a deficiency of vitamin B6, which is also required in the metabolic pathway of tryptophan.

Thiamine is a cofactor in steroid hydroxylation, fatty acid synthesis, and glucose metabolism in the pentose phosphate cycle, which is another energy system. Mercury inhibits glucose-6-phosphatase, a key enzyme in the glucose cycle. The proportion of glucose metabolized by this route is high in the lactating mammary gland, the adrenal cortex, leukocytes and erythrocytes. Each one of the mentioned also are affected by mercury in ways other than glucose metabolism.

Not only mercury, but arsenic and lead impact on the biochemical pathways of these vitamins and minerals. If a person is deficient in thiamine or B1, most likely they are also lacking in all members of the B-complex.

Symptoms Include

- Weakness and fatigue, insomnia, abnormal fears
- Indigestion (hypochlorhydria), dizziness, rage
- Decreased appetite, nervousness, and hostility
- Craving for sweets, instability, depression
- Neuralgia to neuritis, forgetfulness, and anxiety
- Muscular soreness, vague fears, apprehension
- Headache, uneasiness

Vitamin B deficiency leads to heart block and a remarkably slow pulse... Vitamin B deficiency is known to produce a loss of appetite through its effect of causing a loss of tone of the stomach muscles. This loss of tone also may affect the heart and muscular walls of the arterioles--resulting in heart weakness and edema, respectively.

Albert L. Lehninger, in Principles Of Biochemistry, teaches that while muscle cells can oxidize fats for prolonged energy demand and sugar (glucose) for immediate energy...the brain cell has only one Currency For Energy, and that is glucose. Thus, under stress, exaggerated mental activity requires additional glucose for the brain cell to function.

Not only must the brain cell have extra glucose to cope with stress, the cytoplasm and mitochondria of brain cells (neurons), must have all of the members of the B-complex in their biologically-active-form the coenzymes required for anaerobic and aerobic glycolysis.

Minerals and trace elements must also be available optimally. Phosphorus, Potassium, Magnesium, Manganese and Calcium play important roles in the conversion of sugar into energy, within all cells of the body. Being deficient in the B-complex factors, one cannot properly convert sugar or glucose to pyruvic acid for the anaerobic glycolytic pathway of sugar metabolism in the cell cytoplasm. Without B1, pyruvic acid cannot be converted to coenzyme A and enter the mitochondria to participate in the citric acid cycle and form ATP. As a result, pyruvic acid accumulates in nervous tissue and piles up on nerve endings causing irritation. Then the cells signal for help.

The adrenal medula comes to the rescue, increases its function and hormone output...and, as a result, provides the mechanism and the stimulus for producing muscle and brain energy by gluconeogenesis. As long as adrenal hormone output is sufficient, one can survive all the symptoms of B-complex deficiency.

The adrenal glands need protein, minerals and vitamin C (not ascorbic acid), especially tyrosinase (organic copper), to produce catecholamines. Without that extra energy, the person will flee, or cry, or hide. Emotionally they usually hide or separate from reality...the classical nervous breakdown, or nervous exhaustion.

When this inevitable point is near at hand, the person not only needs total nutritional support, but also adrenal support. Besides those nutrients mentioned in the glycolytic phase, the Krebs energy cycle uses B_3 five times, B1 two times, Pantothenic acid two times and Biotin three times. Biotin, normally synthesized in the intestines, is needed to transport CO_2 from the cells into the circulation.

Heavy metal toxicity blocks Biotin function, while antibiotics and chlorine in water, stop or restrict Biotin production in the intestinal tract.

In the anaerobic phase of glycolysis, Fluoride blocks the very important enolase reaction, which impairs glucose metabolism. B2, Riboflavin or"G" combines with several enzymes and enzyme systems forming many flavo-proteins. It is present in retinal pigment contributing to visual acuity.

Deficiency signs and symptoms include:

- Corneal vascularization (red eye)
- Corneal cloudiness and ulceration
- Cataracts
- Photophobia (burning and itching of eyes)
- Abnormal pigmentation of iris
- Stomatitis with lesions
- Glossitis
- Myelin degeneration
- Incoordination
- Loss of strength in arms and legs
- Central neuritis
- Fissures at corners of the mouth
- Scaliness and greasiness at nose & ears
- Impaired red blood cell formation

B_3, or Niacinamide. The acid niacin or nicotinic acid as found in plants is a vasodilator...the amide form as found in animals is not.

B_3 deficiency signs:

- Erythematous cutaneous lesions aggravated by sunlight
- Anorexia, nausea and vomiting
- Dilation of blood vessels (aggravated by Niacin)
- Epithelial atrophy of tongue
- Headaches, dizziness, insomnia, depression, impaired memory
- Enteritis to ulcers
- Fatty infiltration of liver
- Watery to bloody stools
- Myelin degeneration (motor and sensory)
- Failing vision
- Excessive salivation
- Burning hands and feet
- Pain in calves
- Numbness and weakness of extremities
- Difficulty in walking
- Absent knee jerk
- Psychoneurosis to stupor or mania

B_6, or pyridoxine as a pyridoxal-5-phosphate, functions

as a coenzyme in the metabolism of amino acids, fats and fatty acids. Transports amino acids and potassium into cells.

Deficiency signs and symptoms include:

- Seborrheic dermatitis around eyes, eyebrows and angles of mouth
- Glossitis and conjunctivitis
- Severe sensory neuritis
  - Numbness and tingling in hands & feet
  - Hyperesthesia
- Anorexia, nausea, and vomiting
- Lethargy and mental confusion
- Decreased lymphocytes
- Increased Eosinophils
- Elevated B.U.N. (Blood Urea Nitrogen)

FOLIC ACID

Folic acid deficiency is one fo the most common vitamin deficiencies. Many of the symptoms are similar to those of B_{12}deficiency. As much as 100% may be lost if foods are improperly stored, cooking water is discarded, or foods are reheated or overcooked.

Many medications, including aspirin and anticonvulsants, may also interfere with folacin absorption and metabolism, further decreasing its availability. Folic acid is involved in the metabolism of pantothentic acid and is required for its utilization.

It is through this participation that it is involved in antibody production of the immune system. Folic acid also has a relationship with other vitamins including vitamin C and E. In the body, folic acid is rapidly converted to the biologically active form tetrahydrofolic acid (THFA) in the presence of NADPH (niacin's coenzyme form) and ascorbic acid. In conjunction with vitamin B_{12}, THFA participates in amino acid conversions and the methylation of choline, methionine, serine, and histidine.

Folates carry out their metabolic function as a carrier of one-carbon units in the tetra-hydro form. It is this capability to accept one-carbon units, which synthesize various coenzymes, that makes folates so important and involves them in so many metabolic functions. The liver actively reduces and methylates folates which are then transported into the bile for reabsorption by the gut and subsequent delivery to the tissues. This important pathway provides as much as 200 micrograms or more of folate each day for recirculation to tissues. The folates are a family of coenzymes that function in association with their respective enzymes to accomplish many intracellular metabolic functions. Methyl-THFA is required as the methyl donor in the conversion of homocysteine to methionine (methyl methionine).

This reaction utilizes vitamin B_{12} as a cofactor and also requires vitamin B_6. Folates are also involved in the conversion of the amino acid serine to glycine, and histidine to glutamic acid. It is also necessary in the synthesis of glucose, and DNA. Folic acid and vitamin B_{12} are both involved in synthesizing DNA. In megaloblastic anemia (folate deficiency), certain steps in this vital process of DNA synthesis are reduced, causing an arresting of the synthesis of red blood cell replicaiton.

The form of folacin most commonly found in the liver and serum is methyl-THFA. Methyl-THFA can only recycle through a B_{12}-dependent pathway. If a B_{12} deficiency exists, folic acid is trapped as methyl-THFA and is useless to the body. A deficiency of either vitamin B_{12} or folic acid will result in identical hematological symptoms.

A vitamin B_{12} deficiency impairs the conversion of homocysteine to methionine. The enzyme methionine synthetase requires vitamin B_{12}. Therefore, failure of methionine synthesis leads to inadequate formulation of THFA. Methionine deficiency has been equated with B_{12}deficiency. The phagocytic function of polymorphonuclear leucocytes (PMNs), and to a lesser degree bactericidal function, are adversely affected by folic acid deprivation.

Folic acid requirements increase during pregnancy, and low tissue-folate levels are a predisposing factor in the bacteriuria (bacteria in the urine) of pregnancy. There is great potential for mercury and lead to be involved in folate deficiency. The involvement of chronic inhalation of mercury

Vapour from mercury amalgam dental fillings can be an etiological factor in thyroid dysfunction involving folic acid, B_{12}, and thyroxine production.

PANTOTHENIC ACID

The physiological active form of pantothenic acid is coenzyme A, which is required for many different enzymatic actions. It is at the Centre of energy metabolism and fat, acetylcholine, and antibody synthesis.

Coenzyme A, which is synthesized from pantothenic acid, is abbreviated as CoA or CoA-SH, and functions as a transient carrier and a cofactor for a variety of enzyme-catalyzed reactions involving the transfer of acetyl (two carbon) groups. The CoA-SH designation reflects the fact that the coenzyme A molecule has a reactive thiol (-SH) group, to which acyl-groups become covalently linked to formthioesters during acyl-group transfer reactions.

All known acyl derivatives of CoA and related pantetheine derivatives are thiol esters. Because of these derivatives, CoA is involved in the metabolism of carbohydrates, lipids, protein, and porphyrin. It is also involved in the synthesis of fatty acids, cholesterol, citrate, acetoacetate, and the neurotransmitter acetylcholine. Pantothenic acid is an essential component of the brain and must enter the brain and cerebrospinal fluid from the blood. In the brain, as part of CoA, pantothenic acid is involved in many important reactions.

Symptoms identified with pantothenic acid deficiency such as fatigue, headache, insomnia, nausea, abdominal cramps, occasional vomiting, paresthesias of the hands and feet (burning foot syndrome), muscle cramps, and impaired coordination. Pantothenic acid is required for the synthesis of cholesterol precursors for sterol hormones in the adrenal cortex, and a pantothenic deficiency can produce cortical necrosis.

Conversely, pantothenic acid stimulates the adrenal glands and increases production of adrenal hormones. Mercury can inhibit production of adrenal hormones and cause adrenal over-growth. Mercury causes a direct defect in adrenal steroid biosynthesis with physiological consequences being lowered plasma levels of corticosterone and elevated concentrations of progesterone anddehydroepiandrosterone, all abnormal steroid hormone profiles.

Mercury's affinity for thiols, sulfhydryl groups and the sulfur molecule, wherever they are biochemically available in the body, gives it the potential to affect countless metabolic functions. Symptoms produced by a pantothenic deficiency are all also documented as effects produced by mercury toxicity.

Vitamin B_6 (Pyridoxine)

Vitamin B6 participates in the metabolism of protein, carbohydrates, and lipids, and is a coenzyme constituent in the formation of erythrocytes and in amino acid metabolism. The conversion of methionine to cysteine is dependent on vitamin B_6.

The three forms of vitamin B_6 that occur in foods are pyridoxine, pyridoxal phosphate, and pyridoxamine phosphate. The physiologically active forms of vitamin B_6 are pyridoxal phosphate and pyridoxamine phosphate. All three forms are converted in the body to pyridoxal phosphate. Pyridoxal phosphate is involved with a number of enzymes in catalyzing reactions of amino acids. The most common of these are called transaminations, in which an amino group is transferred to a carbon atom.

In these reactions, pyridoxal phosphate serves as a transient carrier of the amino group from its donor to the amino group acceptor. The derivative of this intermediate action is pyridoxamine phosphate which then donates its amino group to the carbon atom of alpha-keto acid, after which it reverts to its pyridoxal phosphate form.

In the biochemical reactions related to the conversion of methionine to cysteine, there is a need to transfer a sulfur component, and an amino component, which would make the process doubly vulnerable to mercury because its highest binding affinity is for the sulfhydryl group and then the amino group. Further, mercury can bind to phosphate groups, which gives it a total of three possible mechanisms to affect the metabolism of pyridoxal phosphate, the biologically active form of vitamin B_6.

One of the functions of vitamin B_6 is called transsulfuration. This is the process whereby one sulfur molecule can replace another and represents an important biochemical function for utilization of sulfur. When pyridoxine is deficient in the body, methionine metabolism is altered. The pancreas is very active in transsulfuration and is known to have a rapid uptake of methionine, because of this, mercury can be involved in pancreatic dysfunction. Cystathionine is a major constituent of the pool of free amino acids in the brain. This fact raises the question of neurochemical involvement. Although vitamin B_6 is not synthesized in the brain, it readily enters the cerebral spinal fluid and brain from the plasma. Once within the CSF, B_6 can enter brain cells.

The holoenzyme needed to convert cystathionine to cysteine contains B_6 in its pyridoxal phosphate form. It is relatively easy to deplete brain B6. Any substance that impedes or inhibits the metabolic functions of vitamin B_6 in humans could have some very serious consequences. Symptoms of vitamin B_6 deficiency, like those of micromercurialism, can be nonspecific and hard to pinpoint.

They include weakness, mental confusion, irritability and nervousness, insomnia, poor coordination in walking, hyperactivity, convulsions, abnormal electroencephalogram, declining blood lymphocytes and white blood cells, anemia, and skin lesions. Mercury has the biochemical capability of reducing the availability of, or inhibiting the function of vitamin B_6. There is a positive correlation between brain mercury levels and the numbers and surfaces of amalgam fillings.

Vitamin B12

Vitamin B_{12} has the most complex structure of all vitamins and is also unique in that it is the only vitamin containing a metal ion. Cobalamin is the generic name of the vitamin because of the presence of the metal ion cobalt in its structure. Vitamin B_{12} was isolated in 1948. Merck and Company is the holder of various patents of procedures for producing synthetic $B_{12}$and owns the product claim. B_{12} functions primarily as an essential coenzyme for the normal metabolism of all cells, including those of the gastrointestinal tract, bone marrow, and nervous tissue.

It is also a coenzyme in the synthesis of red blood cells and in the maintenance of nerve cells, and is involved with protein, lipid and nucleic acid synthesis. It is considered necessary for growth. The coenzyme of vitamin B_{12} is a carrier of methyl groups and hydrogen and is necessary for protein, fat, and carbohydrate metabolism.

Coenzyme B_{12}, methylcobalamin, functions as a methyl-group donor to form methionine from homocysteine. Vitamin B_{12} coenzyme deoxyadenosylco-balamin functions in the conversion of methylmalonic acid to succinic acid. Methionine and B_{12} coordinate in the regulation of folate metabolism. A deficiency of either vitamin results in defective synthesis of DNA in any cell that attempts chromosomal replication and division.

Chronic mercury inhalation from mercury amalgam dental fillings, with its great affinity to bind to methionine and cysteine has the potential to decrease the availability of these amino acids and affect the metabolism of both vitamin B_{12} and folate in man. Methionine is needed in choline synthesis, which means that vitamin B_{12} plays a secondary role in this lipid pathway.

A choline deficiency that causes fatty liver can be prevented by vitamin B_{12} or the other methyl donors--betaine, methionine, and folic acid. Impaired fatty acid synthesis, observed in vitamin B_{12} deficiency conditions, can result in impairment of brain and nerve tissue. The insulation around nerve cells, the myelin sheath, is misformed in vitamin B_{12}deficiency, contributing to faulty nerve transmission.

Neurological disturbances result from prolonged vitamin B_{12} deficiency. Inorganic mercury inhibits outgrowth of nerve Fibres and the development of glial cells, and also depresses the outgrowth of fibroblasts. Methyl-B_{12} is able to inhibit the neurotoxic effects of organic mercury but not those caused by inorganic mercury. Mercury Vapour released from dental amalgam fillings is inorganic mercury. Very low concentrations of mercury and arsenic are able to inhibit nerve outgrowth, and lead directly interferes with the process of myelination.

Proper DNA replication is dependent on the function of coenzyme vitamin B_{12} as a methyl-group carrier. Methyl-B_{12} may play an important role in immune regulation. Methyl-B12 is able to enhance the activity of helper T-cells for immunoglobulin synthesis of B cells. The presence of methyl-B12 significantly potentiates the induction of suppressor cells. Polymorphonuclear leukocytes (PMNs) are white blood cells that function in a phagocytic capacity (capable of destroying foreign substances).

Low levels of mercury inhibit the metabolic reaction called respiratory burst in human PMNs. Mercury acts as an interference to normal biological functions involving the formation and/or function of sulfur containing proteins in cells, enzymes, and hormones. Biological damage will occur long before the appearance of any clinically observable signs and symptoms of mercurialism. Known causes of B_{12} malabsorption are a lack of intrinsic factor and hydrochloric acid in the stomach, the removal or disease of the second portion of the ileum, and competition for B_{12} by microorganisms or parasites.

There is far more unknown than known about B12 and its metabolism in the body. There are several vegetarian food sources of B_{12}. While these sources may not be constant, it appears that they occur with sufficient frequency to supply the minute quantities of B_{12} that are needed by those who are not abusing their health.

These include: wheat, soybeans, various common greens, olives, fruits, and many other foods that occasionally have B_{12} either in or on the food. Vitamin B_{12}has been found in roots and stems of tomatoes, cabbage, celery, kale, broccoli and leeks. It has been found in the leaves of kohlrabi. Seaweed and alfalfa contain B_{12} also. It is difficult to avoid vitamin B_{12}. A further source of B_{12} is bacterial growth in the mouth (around the teeth and gums), in the nasopharynx, around the tonsils and tonsillar crypts, in the folds at the base of the tongue, and in the upper bronchial tree.

Up to 0.5 micorograms daily can be obtained from this source. It's likely that this source alone will supply sufficient quantities of B_{12} for the small requirement that a pure vegetarian has, especially considering his low protein intake, which further reduces the need for B_{12}. It has also been shown that some bacteria, which may colonize the small intestine of man, can synthesize considerable amounts of biologically active forms of the vitamin.

The terminal ileum is principally involved in absorbing B_{12}. Since bacteria can form B_{12} above the terminal ileum, and since intrinsic factor, which is needed for most rapid absorption, is found in the ileum, there can be adequate absorption of vitamin B_{12} produced from these bacteria in the ileum in healthy people.

Individuals taking conventional diets need only about 0.1 micrograms of B_{12} per day even though conventional diets contain excess fat, animal protein and refined foods all of which increase the need for B_{12}. True vegetarians require only 0.05 micrograms of B_{12} daily. An ounce of the roots of leeks, beets and other vegetables will provide 0.1 to 0.3 mcg. of B_1, more than a day's supply. The routine use of vitamin B_{12} is not advised.

Cases of increased cancer production in animals receiving high levels of B_{12} have been published. A group of French investigators reported a series of cases suggesting that B_{12} may stimulate multiplication of cancer cells and aggravate the disease. Patients with rheumatoid arthritis present serum B_{12} levels significantly higher than normal subjects. It is also known to be higher in patients with ulcerative colitis, leukemia and other serious illnesses.

Breast-fed infants get only a tiny fraction of a microgram of B_{12} per day, even if their mothers are on a high intake. Yet, they can build neurological tissue, convert their hemoglobin from F to A type, and do other complex functions dependent on B_{12}. Iron deficiency interferes with absorption of B_{12}. Large doses of vitamin B_1 can also destroy vitamin B_{12}, as can the use of oral contraceptives. Lactose intolerance may increase B_{12} needs.

VITAMIN C COMPLEX

Another danger inherent to crystalline-pure, megavitamin usage is the consumption of ascorbic acid--the so-called vitamin C that is recklessly prescribed by the"pseudo-nutritional pharmacologists" (Physicians, Nurses, Nutrition consultants, registered dieticians, health food store employees, etc.), who have little, if any, realistic training in this most important field. Ascorbic acid is not vitamin C.

Ascorbic acid is a fraction of the biologically utilizable Vitamin C complex. Most all ascorbic acid sold is synthetically produced from corn sugar and even though it is synthetically manufactured, it is called"natural" and"organic" because corn is found in nature. Technically, anything with a carbon atom can be called"organic." The terms natural and organic are totally meaningless.

This article alerts the general public to the fact that as little as 1,500 milligrams of ASCORBIC ACID, taken daily, depletes blood copper, leading to serious health problems. You jeopardize your important"fight or flight" (stress) response by depletion of copper. Just 500 mg. of ascorbic acid with 3 meals can do just that.

Subclinical vitamin C deficiency may exist in a large segment of the population, and can lead to impaired health and increased susceptibility to other diseases. Ascorbic acid therefore is incorrectly named. With isolated ascorbic acid (from Hungarian red pepper), he could not stop the capillary hemorrhaging, so characteristic of scurvy, that he could reliably influence with the C-complex. He went back to the Labouratory and discovered vitamin P, the rutin factor of the C-complex, which exists more in buckwheat (grain and leaf) than in citrus.

The P-factor of the C-complex gives capillary and blood

vessel wall strength, the J-factors, increase the oxygen-carrying capacity of the blood. The known physiological functions of vitamin C are synthesis of polysaccharides and collagen; formation of cartilage, dentin, bone, and teeth; antioxidant; absorption of iron; cold tolerance; maintenance of the adrenal cortex; metabolism of tryptophan, phenylalanine, and tyrosine; growth; wound healing; and maintenance of capillaries. Vitamin C is directly involved in proline and lysine hydroxylation; carnitine synthesis; dopamine hydroxylation; drug and cholesterol breakdown; sulphation; lymphocyte and neutrophil function; and folate reduction.

Vitamin C has a protective effect against mercury poisoning. For many years, mercury diuretics were used extensively by the medical profession. The toxicity of these mercurial diuretics could be reduced if the patient was given vitamin C prior to or simultaneously with the mercurial diuretic. When vitamin C is metabolized, part of it is metabolized to vitamin C-sulphate, with the sulphate being derived from sulphur-containing amino acids such as cysteine. Vitamin C competes with certain drugs for sulphate conjugation, which could affect the pharmacological activity and toxicity of drugs.

Mercury competes actively for the sulphur-containing amino acids cysteine and methionine, as well as the cysteine molecule of the tripeptide glutathione. Anyone with amalgam fillings should routinely eat a diet rich in sulphur-containing foods to offset the depletion of cysteine and methionine caused by the inhalation of mercury Vapour being continually released from amalgam fillings. The problem of cysteine depletion is further aggravated by the presence of mercury, lead, arsenic, and cadmium, because they are all thiol sensitive.

Mercury is also known to inhibit collagen synthesis. The synthesis of collagen is impaired in vitamin C deficiency, due to lowered ability to hydroxylate lysine and proline. The reducing agent in the hydroxylation of both lysine and proline is the reduced form of vitamin C. Consequently, in vitamin C deficiency, the amount of effective collagen Fibre present in connective tissue is reduced. In severely scorbutic patients, swelling, hemorrhages, and secondary bacterial infections of the gingival margins are common.

It is not the actual deficiency of vitamin C that causes the inflammation, but the lack of real vitamin C, impairing the normal defensive responses of the mucous membranes. Thus, the massive gingival enlargement so characteristic of scurvy results from the combined effects of lack of vitamin C and nonspecific inflammation. The adrenal glands are one of the target glands of mercury deposition. They also contain the body's second highest tissue levels of vitamin C.

The physical response to stress is an increased secretion of the hormones of the adrenal glands. Physical and mental stress increases adrenal activity which, in turn, depletes vitamin C from the gland. Humans replenish the adrenal stores of vitamin C by taking it from other stores in the body.

If tissue values of vitamin C are low, there may be an insufficient amount available to replenish or satisfy the requirements of the adrenals. Thiol groups have a direct role in the preservation of tissue vitamin C and its reduced form, and any decrease in the biological reducing capacity of tissue thiols will result in lower levels of tisssue vitamin C. Under these conditions, normal adrenal hormone response may become inadequate. Prolonged exposures to low mercury concentrations depress the adrenal vitamin C content.

The person with amalgam dental fillings who is being exposed to chronic intakes of mercury Vapour is subjecting the adrenals to depletion by chemical stress. One function of the liver tissue is to reduce dehydroascorbic acid to ascorbic acid. This ability guarantees the preservation of the body's vitamin C reserves, so necessary to normal function.

The body doesn't want ascorbic acid *per se;* it wants the C-complex being preserved by ascorbic acid. Once in the body, the body sheds the ascorbic acid, the preservative wrapper, the way you shed the peel of an orange; takes the rest of the C-complex and places it as the lance of the lymphocyte; and kills bacteria with it. A lymphocyte unarmed with C complex will fail in its function, it will be impotent, it won't be able to destroy infectious organisms. Lymphocyte potency is dependent on copper, the core trace mineral of the tyrosinase enzyme.

The reason one gets any relief from taking ascorbic acid is that ascorbic acid lowers the pH to the acid side of the pH scale. It acidifies the body, creating an unfavourable medium of growth for pathogenic bacteria. Acting in this way, ascorbic acid is not producing a"vitamin effect" but rather an"acidifying effect." Most infectious pathogenic bacteria thrive in an alkaline pH.

The following bacteria, all well-known enemies of modern science's war on bacterial infection, grew optimally on alkaline media of pH 7.4 and above: staphylococcus (staph infection), streptococcus (strep throat), pneumococcus (pneumonia), h. influenza (the flu), meningococcus (meningitis), corymbacterium diptheria (diptheria), clostridium tetani (tetanus), and others.

Apple cider vinegar is 5% malic acid, a natural Colourless crystalline acid normally found in the body. It can be taken with water during meals, and is much less expensive than ascorbic acid and does a better job in lower doses at acidifying the body, without causing rebound deficiencies as ascorbic acid does. Ascorbic acid has long been recognized as a contributing factor in tyrosine (copper) depletion. Tyrosinase is absolutely essential in the biosynthesis of catecholamines.

The adrenal medulla supplies the catecholamines, epinephrine and norepinephrine which are biosynthesized from phenylalanine being oxidized to tyrosine, to dopa (dihydroxyphenylalanine), dopa is transformed first into dopamine and then into norepinephrine with the enzyme tyrosinase catalyzing the intermediate reaction. Other functions of organic copper include: vascular integrity, purine excretion, hemoglobin stability, gas exchange, digestive enzymes, myelin maintenance, visual accommodation, fertility, adrenal hormone synthesis, and connective tissue integrity.

The former, catalase, is an enzyme that cooperates with SOD in the removal of free oxygen radicals from cellular metabolism. Excess quantities of so-called vitamin C can reduce the pH of the urine to as low as 4.0, and cause the precipitation of large quantities of urates, increasing incidence of kidney stones. It appears that excessive doses of ascorbic acid interfere with purine metabolism, increasing the risk of gout. There is some evidence that high doses of ascorbic acid reduce fertility in some women. Large doses of ascorbic acid and vitamin E are undesirable as they may reduce availability of vitamin A.

Ascorbic acid increases intestinal peristalsis and may produce diarrhea. This laxative action, along with its acidifying effect, may be the only benefit ascorbic acid has against a cold. High doses of so-called vitamin C, as popularly used in home remedies for the common cold, destroy substantial amounts of vitamin B12.

Ascorbic acid in quantities in excess of a half a gram will destroy 50% to 95% of B12 content in food. Megadoses of phony vitamin C may produce B12 deficiency by destroying the cobalamins during transport through the digestive tract, and possibly also in the tissues. No further discussion should be needed to establish the lack of wisdom leading to taking or prescribing ascorbic acid.

Vitamin E

The chemical name for vitamin E is tocopherol, which is derived from the Greek tokos (childbirth) and pherin (to bear). The ending ol is the chemical suffix to denote an alcohol. The name tocopherol was bestowed on this vitamin in 1938. Vitamin E is a fat-soluble vitamin, as are vitamins A, D, and K. Fat-soluble vitamins can be stored in the body whereas water-solubles cannot. Fat-soluble vitamins are excreted chiefly by the fecal pathway versus the urinary pathway for the water-soluble vitamins. They are absorbed along with dietary fats, and conditions of extremely low fat intake or impaired uptake of fats will also interfere with their absorption. Antibiotics and certain other drugs, as well as certain disease states such as malabsorption syndromes, decrease the absorption of fat-soluble vitamins from the intestinal tract.

One of the primary biological functions of vitamin E relates to its role as an antioxidant. Vitamin E's antioxidant ability is enhanced by selenium. This synergistic effect preserves membranes from destruction by oxidation products and especially retards breakdown (hemolysis) of red blood cells. Vitamin E can reduce the toxic effects of mercury. Vitamin E is able to reduce the chromosomal breakage caused by mercury. Vitamin E also has sulfhydryl-protective activity.

When vitamin E is involved in reducing the toxic effects of mercury, there is less available for normal metabolic functions. Vitamin E protects the enzymes G6P, G6PD, GSH-Px and glutathione reductase from auto-oxidative damage and at the same time protects the biological supply of glutathione. The presence of mercury in combination with a vitamin E deficiency has the potential of producing toxic reactions in the body. The primary lymphoid organs are major targets of selenium and vitamin E dietary deficiencies, contributing to impaired immune function.

Vitamin E is not a cofactor, such as the trace elements manganese, copper, zinc, and selenium, it is an independent antioxidant that has its main effect in the lipid phase of the cell, protecting the membrane from auto-oxidative damage. A distinguishing feature of cystic fibrosis is fat malabsorption. Cystic fibrosis patients usually have a deficiency of vitamin E. Patients with G6PD deficiency have less capacity to generate reduced glutathione, and their red blood cells have a greatly reduced half-life. Vitamin E supplementaton can decrease the level of neonatalhyperbilirubinemia due to G6PD deficiency. Mercury inhibits G6PD activity and also reduces available vitamin E.

Alpha and delta tocopherols or mixed tocopherols are not vitamin E. They are merely antioxidant or protective agents (banana peel) for the more complex parts of the vitamin E assemblage (xanthine, phospholipids, lipositols, sex hormone precursors)--see functional architecture of vitamin E. There are seven tocopherols in all, alpha being one of them. The government has decided to rate any vitamin E product according to the amount of alpha tocopherol it contains.

The tocopherols are antioxidants, just like ascorbic acid. They are a part of the E complex, but they are not the active ingredient.

Their function is to preserve the active factor. Toccpherols are measured on the basis of cell growth. The entire E complex contains the polyunsaturated fatty acids. (Vitamin F) Vitamin F is part of the E complex. Also, vitamins A and K. There are some forms of vitamin D and manganese in the E complex, as well. The traditional source for vitamin E is vegetable oils. But the best quality vitamin E comes from vegetables.

Lettuce is one of the best sources and the pea plant is equally as good. If you take any Labouratory-chelated products, your body has to do some combining to make them work. But if you take your minerals the way nature provides them, combined with protein and vitamins through the growing process of plants, they are bound to work. For example, we could take vitamin E and separate the tocopherols from the manganese.

Putting them back together in the Labouratory--chelating them artificially--we would find they wouldn't work like they did before you took them apart. If you take a watch apart piece by piece and Analyse it and then throw all the parts back together, it isn't going to work! Although it is the same watch, it has now lost its function. A nutritional concentrate is an organized mechanism that has a function. If you break the function by separating the parts, the body can no longer use the mechanism.

High potency vitamin E is made this way. Tocopherols are broken down and separated from oils--the natural tocopherols are separated from vegetable oil.

Then they are packed into a gelatin perle, and you have a high potency vitamin E product--a 220 or 500 International Unit (IU) Vitamin E product. That means 200 or 500 International Units) of alpha tocohpherol--the potency of the product is measured by the amount of alpha tocopherol it contains.

Although the label of such a product will say"all natural" or"organic""--it was taken from oil, a natural source--it's still not food. It's an incomplete product. You're getting lots of tocopherols, but none of the active components of the E complex; the F, A, K, D, manganese, selenium, etc.

The Vitamin E2fraction (prostaglandin) is found in beef chromatin material, which is glandular substance. It's a food factor that will stop a heart cramp or a heart pain, usually within about ten minutes.

Most cramps are due to calcium deficiency. So E2 gets calcium into the area to relax the muscle. It can be used to relieve any muscle cramp, but it's especially effective to relax cramps in the muscles of the heart.

Natural forms of vitamin E complex lose up to 99% of their potency when separated from their natural synergists (Ann. Review Biochemistry, 1943, page 381). These synergists include tannins, fatty acids (vitamin F complex), phospholipids, and other synergistic factors. Chemically purified vitamin E (tocopherols), in high unit doses, reverses its effect and produces the same symptoms as a deficiency (including bone decalcification).

Without selenium (the trace mineral activator of vitamin E), vitamin E cannot perform many of its functions.

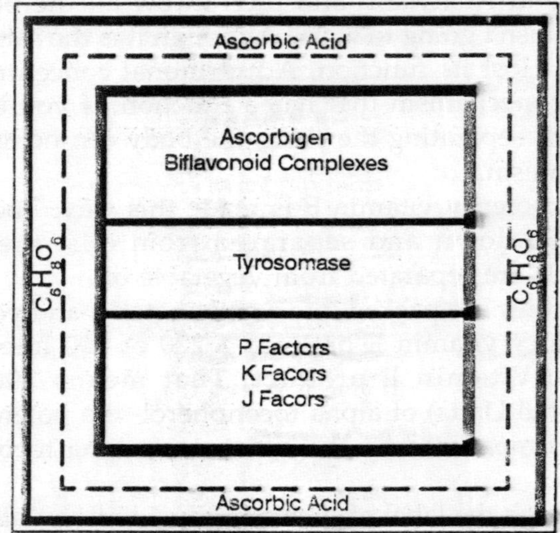

Fig. Function Architecture of Vitamin C-Complex

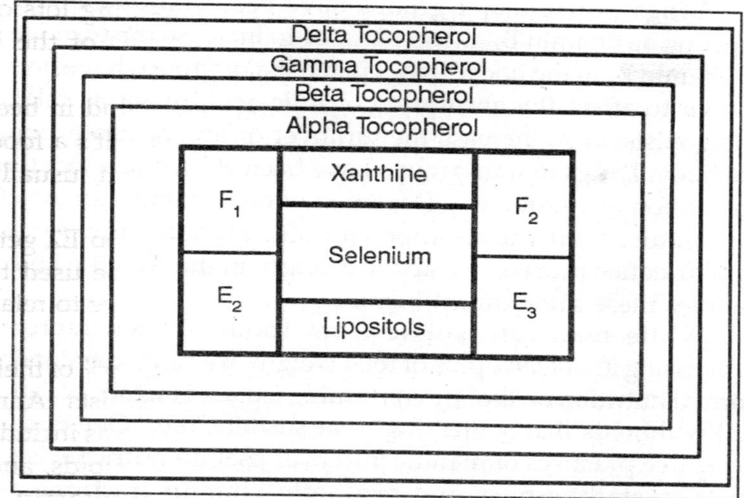

Fig. Functional Architecture of Vitamin E-Complex

Xanthine increases the antioxidative function of the tocopherols as much as 50%. Xanthine plays an important role in the metabolism of purines (nucleoproteins) within the cell. In one instance, hypoxanthine is transformed biochemically to xanthine with the metabolic fate proceeding to uric acid in two forms; enol and keto. Intracellular utilization of purines is dependent upon the said contribution. Lipositols are essential to cholesterol metabolism in order that cholesterol be altered to a biochemical acceptable to the synthesis of adrenal and sex hormones. F1 is the active form of unsaturated fatty acids described in the Annual Review of Biochemistry, 1949. Vitamin F restores and maintains Calcium in tissue fluids and is anticarcinogenic. F_2 is part of the phospholipid complex in nerve structures; particularly the myelin sheath. It also promotes tissue repair and enhances the benefits of the antiarthritic Wulzen factor. F_2 could very well be the missing link in pernicious anemia according to Dr. Royal Lee. E_2 is a part of E complex that has been compared to nitroglycerine in its function of relieving angina-type pains--the hypoxic spasm of the heart muscle. E_2 reduces pain because it increases the oxygen conserving factor of the blood. The E_3 component of Vitamin E complex is a precursor of sex hormones.

Drugs in five major groups have been shown either to function as vitamin B_6 antagonists, or to increase the turnover of vitamin B_6 in the body. Drugs in ten major groups have been shown to affect the absorption of folic acid; to act as folate antagonists; or to increase the turnover or loss of folate from the body. Drugs in four groups have been shown to affect the absorption of vitamin B_{12}. Depletion of one vitamin can affect the requirement for another vitamin. Mercury can cause similar deficiencies or metabolic changes in the way the body handles these same three vitamins.

While man can isolate from foods known factors comprising about 99% plus of total weight, we cannot maintain normal nutrition or"life" by consuming isolated factors. Nature still withholds that"something" that sustains"life." One of the deceptive practices of manufacturers of pseudo-nutrients is to label crystalline-pure and synthetic vitamin fractions as COMPLEXES. The meaning of complex is simply: the sum or combination of various things. Another, but more precise, definition is: Consisting of various connected or interwoven parts; composite. As a result, the consumer buys a"vitamin" product that says on the label:"Vitamins From Natural Sources (usually synthetic, from corn sugar and coal tar) with Bioflavonoids or Rutin." Legally, this can be called a"Vitamin Complex" but it is really no more than a mixture of two or more dead, refined chemicals--not a required for life,"biochemical." Remember, the human anatomy is composed of 100 trillion cells that need 75+ nutrients every day and, in turn, it can require six to eight months or more to overcome a nutritional deficiency. If you try and fool mother nature, she'll make a fool out of you.

READING VITAMIN LABELS

NATURAL in this writing means,"As naturally found in nature or natural food sources." It further implies that in preparation there has been no change, either biologically or chemically, in the combination or in the activity of these natural foods. This does not mean the moisture and Fibre cannot be removed.

CO-NATURAL or CRYSTALLINE A co-natural vitamin is one that has a natural food as its source, but has been reduced to one specific pure crystalline vitamin by means of solvents, heat or distillation. Since it is practically impossible to reduce any natural food product to an absolutely pure state during manufacture, the impurities remaining could be certain synergists.

A synthetic vitamin is created by a scientist in a Labouratory when he/she has reconstructed the exact molecular structure of a vitamin by putting together or chemically combining molecules from other know sources. The synthetic vitamin has no synergists, no enzymes or co-enzymes, minerals, mineral activators, or co-vitamin helpers. It is a chemically pure vitamin but is, conformationally, a mirror image of the real thing. Legally it is not necessary to give the source from which the synthetic chemical is derived (coal-tar).

Item	**If Source is**	**It is**
Vitamin A	Fish oils Acetate Palmitate Lemongrass	Natural Synthetic Synthetic Co-Natural
	Vitamin B Complex	
Vitamin B_1 (Thiamine)	Yeast (unenriched) Thiamine mononitrate Thiamine hydrochloride	Natural Synthetic Synthetic
Vitamin B_2 (riboflavin)	Yeast (unenriched)	Natural
Pantothenic Acid	Yeast (unenriched) rice bran, liver Cal panto	Natural Synthetic
Vitamin B_3 (Niacin)	Yeast (unenriched), or bran Niacinamide	Natural Co-Natural
Vitamin B_6 (Pyridoxine)	Yeast (unenriched) Pyridoxine hydrochloride	Natural Synthetic

Vitamin B_{12}	Liver or Yeast (unenriched)	Natural
	Cobalamin from fermentation	Natural
	Cyano cobalamin	Co-Natural/ Crystalline
	Cyano cobalamin	Co-Natural/ Crystalline
PABA	Yeast (unenriched)	Natural
	Para-aminobenzoic acid	Synthetic
Folic Acid	Yeast (unenriched) or Liver	Natural
	Pteroylglutamic acid	Synthetic
Inositol	Soy beans	Natural
	Reduced from corn	Co-Natural
Choline	Soy beans	Natural
	Choline chloride	Synthetic
	Choline bitartrate	Synthetic
Biotin	Liver	Natural
	d-Biotin	Synthetic
Vitamin B_{15} (Pangamic acid)	Bran-apricot kernels	Natural
	Pangamic acid	Synthetic
Vitamin C-Complex		
Vitamin C	Buckwheat leaf, peppers, tomatoes, citrus, rose hips	Natural
	Acerola berries	Co-Natural
	Ascorbic acid	Synthetic
Vitamin D-Complex		
Vitamin D	Fish oil	Natural
	Irradiated ergosterol	Co-Natural
	Calciferol	Synthetic
Vitamin E-Complex		
Vitamin E	Wheat germ oil	Natural
	dl-alpha tocopherol	Synthetic

	d-alpha tocopherol acetate d-alpha tocopherol succinate	Synthetic Synthetic
Vitamin F		
(Unsaturated fatty acids)	Flax oil, pumpkin seeds, walnuts, soybeans	Natural
Vitamin K		
Vitamin K	Alfalfa, Cabbage Menadione	Natural Synthetic

Folic Acid (Pteroylglutamic Acid)

Folic acid is derived from the addition of a pteroyl group to the amino acid glutamate. Folic acid can be reduced to form tetrahydrofolate, which is an important acceptor of one-carbon units. Folic acid plays an important role in the conversion of homocysteine to methionine by providing the methyl to the vitamin B,2 dependent methyl transferase. Therefore, folic acid levels have a strong influence on the endogenous concentration of homocysteine. Elevated levels of homocysteine are associated with increased risk of coronary heart disease. Supplementation with folic acid has been shown to correct plasma homocysteine levels and decrease the morbidity and mortality from atherosclerotic disease.

Chapter 5

Animal and Plant Hormones

PLANT-HORMONES

The plant hormones website was set up initially at Long Ashton Research Station an institute of the BBSRC. Long Ashton will be closing in March 2003 so the Plant-Hormones site has been moved to a new commercial service. The website provides general information on plant hormones and links to the plant-hormones discussion list. This list has been set up on the National Academic Mailing List Service JISCMAIL.

This list has been set up as a freely accessible electronic communication forum for use by plant-hormone scientists. The list is intended to promote communication between professionals in the plant-hormone field.

Discussion topics should include gibberellins, auxins, cytokinins,abscisic acid, etc. All aspects of research may be discussed; synthesis extraction, purification, analysis, enzymology, molecular biology, hormone action etc. Requests for help on particular aspects may also be posted to the list as well as requests for standards, spectra, identification of spectra etc. The list may also be used for announcement of meetings, calls for papers and posters, visits of scientists to a particular area.

ABSCISIC ACID

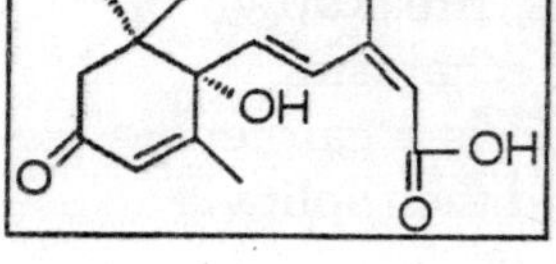

Nature of Abscisic Acid

Abscisic acid is a single compound unlike the auxins, gibberellins, and cytokinins. It was called"abscisin II" originally because it was thought to play a major role in abscission of fruits. At about the same time another group was calling it"dormin" because they thought it had a major role in bud dormancy. The name abscisic acid (ABA) was coined by a compromise between the two groups. Though ABA generally is thought to play mostly inhibitory roles, it has many promoting functions as well.

History of Abscisic Acid

In 1963, abscisic acid was first identified and characterized by Frederick Addicott and his associates. They were studying compounds responsible for the abscission of fruits (cotton). Two compounds were isolated and called abscisin I and abscisin II. Abscisin II is presently called abscisic acid (ABA). Two other groups at about the same time discovered the same compound. One group headed by Philip Wareing was studying bud dormancy in woody plants. The other group led by Van Steveninck was studying abscission of flowers and fruits from lupine. Plant physiologists agreed to call the compound abscisic acid.

Biosynthesis and Metabolism

ABA is a naturally occurring compound in plants. It is a sesquiterpenoid (15-carbon) which is partially produced via the mevalonic pathway in chloroplasts and other plastids. Because it is sythesized partially in the chloroplasts, it makes sense that biosynthesis primarily occurs in the leaves. The production of ABA is accentuated by stresses such as water loss and freezing temperatures. It is believed that biosynthesis occurs indirectly through the production of carotenoids. Carotenoids are pigments produced by the chloroplast which have 40 carbons. Breakdown of these carotenoids occurs by the following mechanism:

Violaxanthin is a carotenoid which has forty carbons. It is isomerized and then split via an isomerase reaction followed by an oxidation reaction. One molecule of xanthonin is produced from one molecule of violaxanthonin and it is uncertain what happens to the remaining biproduct. The one molecule of xanthonin produced is unstable and spontaneously changed to ABA aldehyde. Further oxidation results in ABA.

Activation of the molecule can occur by two methods. In the first method, an ABA-glucose ester can form by attachment of glucose to ABA. In the second method, oxidation of ABA can occur to form phaseic acid and dihyhdrophaseic acid. The transport of ABA can occur in both xylem and phloem tissues. It can also be translocated through paranchyma cells. The movement of abscisic acid in plants does not exhibit polarity like auxins. ABA is capable of moving both up and down the stem.

Functions of Abscisic Acid

The following are some of the phyysiological responses known to be associated with abscisic acid

- Stimulates the closure of stomata (water stress brings about an increase in ABA synthesis).
- Inhibits shoot growth but will not have as much affect on roots or may even promote growth of roots.
- Induces seeds to synthesize storage proteins.
- Inhibits the affect of gibberellins on stimulating de novo synthesis of a-amylase.
- Has some effect on induction and maintanance of dormancy.

Induces gene transcription especially for proteinase inhibitors in response to wounding which may explain an apparent role in pathogen Defence.

AUXINS

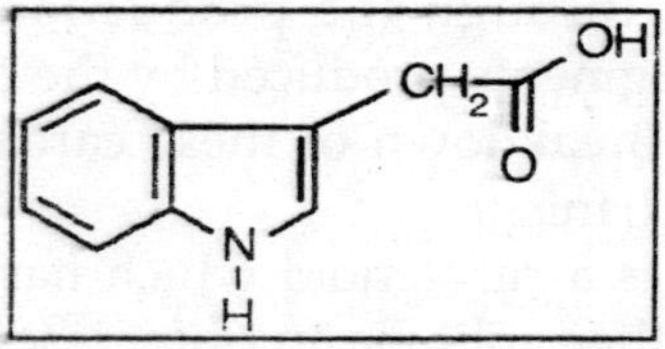

Nature of Auxins

The term auxin is derived from the Greek word auxein which means to grow. Compounds are generally considered auxins if they can be characterized by their ability to induce cell elongation in stems and otherwise resemble indoleacetic acid (the first auxin isolated) in physiological activity. Auxins usually affect other processes in addition to cell elongation of stem cells but this characteristic is considered critical of all auxins and thus"helps" define the hormone.

History of Auxins and Pioneering Experiments

Auxins were the first plant hormones discovered. Charles Darwin was among the first scientists to dabble in plant hormone research. In his book"The Power of Movement in Plants" presented in 1880, he first describes the effects of light on movement of canary grass coleoptiles. The coleoptile is a specialized leaf originating from the first node which sheaths the epicotyl in the plants seedling stage protecting it until it emerges from the ground.

When unidirectional light shines on the coleoptile, it bends in the direction of the light. If the tip of the coleoptile was covered with aluminum foil, no bending would occur towards the unidirectional light. However if the tip of the coleoptile was left uncovered but the portion just below the tip was covered, exposure to unidirectional light resulted in curvature towards the light. Darwin's experiment suggested that the tip of the coleoptile was the tissue responsible for perceiving the light and producing some signal which was transported to the lower part of the coleoptile where the physiological response of bending occurred.

He then cut off the tip of the coleoptile and exposed the rest of the coleoptile to unidirectional light to see if curving occurred. Curvature did not occur confirming the results of his first experiment.

It was in 1885 that Salkowski discovered indole-3-acetic acid (IAA) in fermentation media. The isolation of the same product from plant tissues would not be found in plant tissues for almost 50 years. IAA is the major auxin involved in many of the physiological processes in plants. In 1907, Fitting studied the effect of making incisions on either the light or dark side of the plant.

His results were aimed at understanding if translocation of the signal occurred on a particular side of the plant but his results were inconclusive because the signal was capable of crossing or going around the incision.

In 1913, Boysen-Jensen modified Fritting's experiment by inserting pieces of mica to block the transport of the signal and showed that transport of auxin towards the base occurs on the dark side of the plant as opposed to the side exposed to the unidirectional light. In 1918, Paal confirmed Boysen-Jensen's results by cutting off coleoptile tips in the dark, exposing only the tips to the light, replacing the coleoptile tips on the plant but off Centreed to one side or the other.

Results showed that whichever side was exposed to the coleoptile, curvature occurred towards the other side. Soding was the next scientist to extend auxin research by extending on Paal's idea. He showed that if tips were cut off there was a reduction in growth but if they were cut off and then replaced growth continued to occur.

In 1926, a graduate student from Holland by the name of Fritz Went published a report describing how he isolated a plant growth substance by placing agar blocks under coleoptile tips for a period of time then removing them and placing them on decapitated Avena stems. After placement of the agar, the stems resumed growth

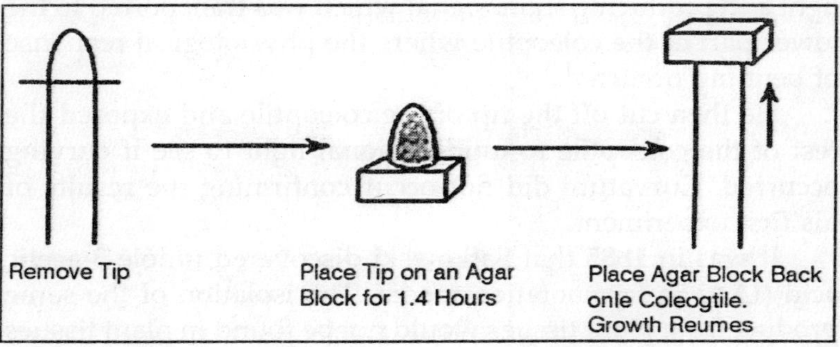

In 1928, Went developed a method of quantifying this plant growth substance. His results suggested that the curvatures of stems were proportional to the amount of growth substance in the agar.

This test was called the avena curvature test.(see below) Much of our current knowledge of auxin was obtained from its applications. Went's work had a great influence in stimulating plant growth substance research. He is often credited with dubbing the term auxin but it was actually Kogl and Haagen-Smit who purified the compound auxentriolic acid (auxin A) from human urine in 1931. Later Kogl isolated other compounds from urine which were similar in structure and function to auxin A, one of which was indole-3 acetic acid (IAA) initially discovered by Salkowski in 1985. In 1954 a committee of plant physiologists was set up to characterize the group auxins. The term comes from the Greek auxein meaning"to grow." Compounds are generally considered auxins if they are synthesized by the plant and are substances which share similar activity to IAA.

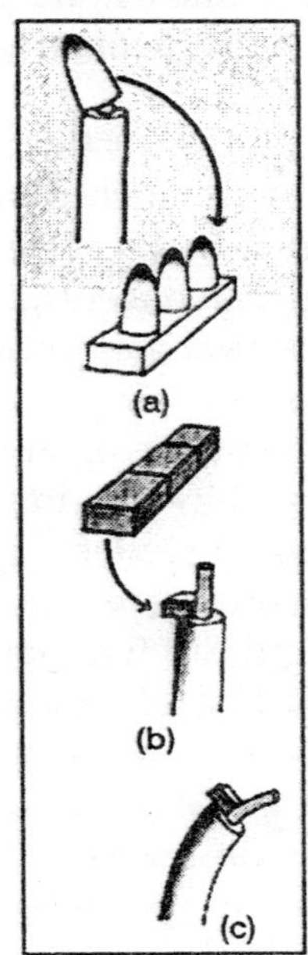

BIOSYNTHESIS AND METABOLISM OF AUXIN

IAA is chemically similar to the amino acid tryptophan which is generally accepted to be the molecule from which IAA is derived. Three mechanisms have been suggested to explain this conversion:

Tryptophan is converted to indolepyruvic acid through a transamination reaction. Indolepyruvic acid is then converted to indoleacetaldehyde by a decarboxylation reaction. The final step involves oxidation of indoleacetaldehyde resulting in indoleacetic acid.

Tryptophan undergoes decarboxylation resulting in tryptamine. Tryptamine is then oxidized and deaminated to produce indoleacetaldehyde. This molecule is further oxidized to produce indoleacetic acid. As recently as 1991, this 3rd mechanism has evolved. IAA can be produced via a tryptophan-independent mechanism. This mechanism is poorly understood, but has been proven using trp(-) mutants. Other experiments have shown that, in some plants, this mechanism is actually the preferred mechanism of IAA biosynthesis.

The enzymes responsible for the biosynthesis of IAA are most active in young tissues such as shoot apical meristems and growing leaves and fruits. The same tissues are the locations where the highest concentrations of IAA are found. One way plants can control the amount of IAA present in tissues at a particular time is by controlling the biosynthesis of the hormone. Another control mechanism involves the production of conjugates which are, in simple terms, molecules which resemble the hormone but are inactive.

The formation of conjugates may be a mechanism of storing and transporting the active hormone. Conjugates can be formed from IAA via hydrolase enzymes. Conjugates can be rapidly activated by environmental stimuli signaling a quick hormonal response. Degradation of auxin is the final method of controlling auxin levels.

This process also has two proposed mechanisms outlined below: The oxidation of IAA by oxygen resulting in the loss of the carboxyl group and 3-methyleneoxindole as the major breakdown product. IAA oxidase is the enzyme which catalyzes this activity. Conjugates of IAA and synthetic auxins such as 2,4-D can not be destroyed by this activity. C-2 of the heterocyclic ring may be oxidized resulting in oxindole-3-acetic acid. C-3 may be oxidized in addition to C-2 resulting in dioxindole-3-acetic acid.

The mechanisms by which biosynthesis and degradation of auxin molecules occur are important to future agricultural applications. Information regarding auxin metabolism will most likely lead to genetic and chemical manipulation of endogenous hormone levels resulting in desirable growth and differentiation of important crop species. Ultimately, the possibility exists to regulate plant growth without the use of hazardous herbicides and fertilizers.

Functions of Auxin

The following are some of the responses that auxin is known to cause (Davies, 1995; Mauseth, 1991; Raven, 1992; Salisbury and Ross, 1992).

- Stimulates cell elongation
- Stimulates cell division in the cambium and, in combination with cytokinins in tissue culture
- Stimulates differentiation of phloem and xylem.
- Stimulates root initiation on stem cuttings and lateral root development in tissue culture.
- Mediates the tropistic response of bending in response to gravity and light.
- The auxin supply from the apical bud suppresses growth of lateral buds.
- Delays leaf senescence.
- Can inhibit or promote (via ethylene stimulation) leaf and fruit abscission.
- Can induce fruit setting and growth in some plants.
- Involved in assimilate movement towards auxin. possibly by an effect on phloem transport.
- Delays fruit ripening.
- Promotes flowering in Bromeliads.

- Stimulates growth of flower parts.
- Promotes (via ethylene production) femaleness in dioecious flowers.
- Stimulates the production of ethylene at high concentrations.

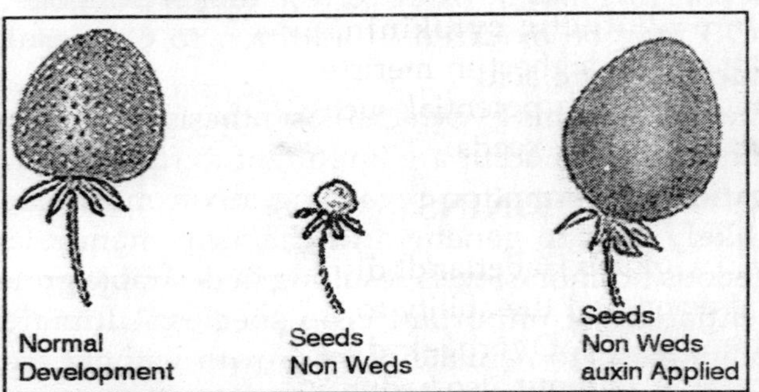

Above describes the effect of auxin on strawberry development. The achenes produce auxin. When removed the strawberry does not develop.

CYTOKININS

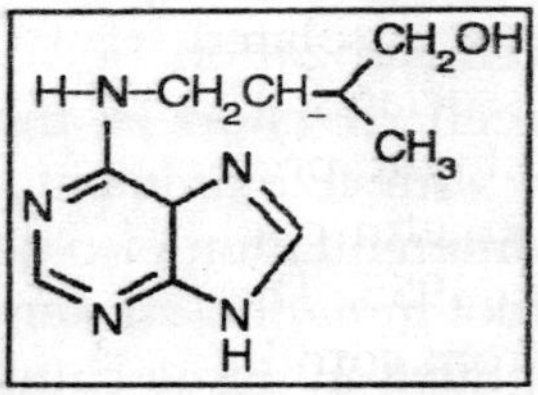

CH_2OH

H–N–CH_2CH–

CH_3

N N N N H

NATURE OF CYTOKININS

Cytokinins are compounds with a structure resembling adenine which promote cell division and have other similar functions to kinetin.

Kinetin was the first cytokinin discovered and so named because of the compounds ability to promote cytokinesis (cell division).

Though it is a natural compound, It is not made in plants, and is therefore usually considered a"synthetic" cytokinin (meaning that the hormone is synthesized somewhere other than in a plant). The most common form of naturally occurring cytokinin in plants today is called zeatin which was isolated from corn (Zea mays).

Cytokinins have been found in almost all higher plants as well as mosses, fungi, bacteria, and also in tRNA of many prokaryotes and eukaryotes. Today there are more than 200 natural and synthetic cytokinins combined. Cytokinin concentrations are highest in meristematic regions and areas of continuous growth potential such as roots, young leaves, developing fruits, and seeds.

HISTORY OF CYTOKININS

In 1913, Gottlieb Haberlandt discovered that a compound found in phloem had the ability to stimulate cell division. In 1941, Johannes van Overbeek discovered that the milky endosperm from coconut also had this ability. He also showed that various other plant species had compounds which stimulated cell division. In 1954, Jablonski and Skoog extended the work of Haberlandt showing that vascular tissues contained compounds which promote cell division (Jablonski and Skoog, 1954).

The first cytokinin was isolated from herring sperm in 1955 by Miller and his associates. This compound was named kinetin because of its ability to promote cytokinesis. Hall and deRopp reported that kinetin could be formed from DNA degradation products in 1955. The first naturally occurring cytokinin was isolated from corn in 1961 by Miller.

It was later called zeatin. Almost simultaneous with Miller Letham published a report on zeatin as a factor inducing cell division and later described its chemical properties. It is Miller and Letham that are credited with the simultaneous discovery of zeatin. Since that time, many more naturally occurring cytokinins have been isolated and the compound is ubiquitous to all plant species in one form or another.

BIOSYNTHESIS AND METABOLISM OF CYTOKININS

Cytokinin is generally found in higher concentrations in meristematic regions and growing tissues. They are believed to be synthesized in the roots and translocated via the xylem to shoots. Cytokinin biosynthesis happens through the biochemical modification of adenine. The process by which they are synthesized is as follows:

- A product of the mevalonate pathway called isopentyl pyrophosphate is isomerized.
- This isomer can then react with adenosine monophosphate with the aid of an enzyme called isopentenyl AMP synthase.
- The result is isopentenyl adenosine-5'-phosphate (isopentenyl AMP).
- This product can then be converted to isopentenyl adenosine by removal of the phosphate by a phosphatase and further converted to isopentenyl adenine by removal of the ribose group.
- Isopentenyl adenine can be converted to the three major forms of naturally occurring cytokinins.
- Other pathways or slight alterations of this one probably lead to the other forms.

Degradation of cytokinins occurs largely due to the enzyme cytokinin oxidase. This enzyme removes the side chain and releases adenine. Derivitives can also be made but the pathways are more complex and poorly understood.

CYTOKININ FUNCTIONS

A list of some of the known physiological effects caused by cytokinins are listed below. The response will vary depending on the type of cytokinin and plant species.

- Stimulates cell division.
- Stimulates morphogenesis (shoot initiation/bud formation) in tissue culture.
- Stimulates the growth of lateral buds-release of apical dominance.
- Stimulates leaf expansion resulting from cell enlargement.
- May enhance stomatal opening in some species.
- Promotes the conversion of etioplasts into chloroplasts via stimulation of chlorophyll synthesis.

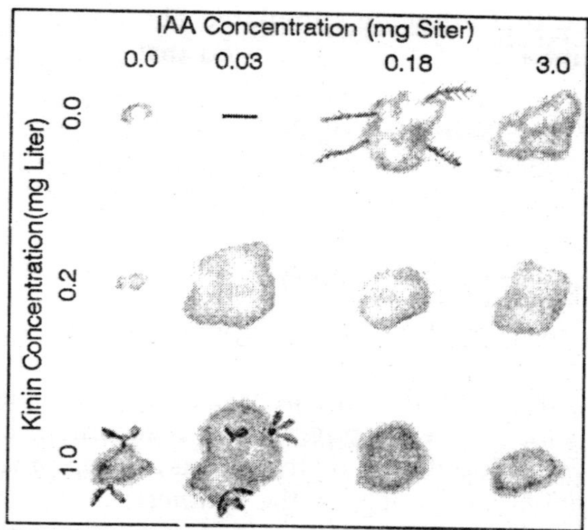

The illustration above shows the effect of cytokinin and auxin concentration on tissue culture experiments.

ETHYLENE

$$H_2C=CH_2$$

NATURE OF ETHYLENE

Ethylene, unlike the rest of the plant hormone compounds is a gaseous hormone. Like abscisic acid, it is the only member of its class. Of all the known plant growth substance, ethylene has the simplest structure. It is produced in all higher plants and is usually associated with fruit ripening and the tripple response.

HISTORY OF PLANTS

Ethylene has been used in practice since the ancient Egyptians, who would gas figs in order to stimulate ripening. The ancient Chinese would burn incense in closed rooms to enhance the ripening of pears. It was in 1864, that leaks of gas from street lights showed stunting of growth, twisting of plants, and abnormal thickening of stems (the triple response).

In 1901, a russian scientist named Dimitry Neljubow showed that the active component was ethylene. Doubt discovered that ethylene stimulated abscission in 1917. It wasn't until 1934 that Gane reported that plants synthesize ethylene.

In 1935, Crocker proposed that ethylene was the plant hormone responsible for fruit ripening as well as inhibition of vegetative tissues. Ethylene is now known to have many other functions as well.

Biosynthesis and Metabolism

Ethylene is produced in all higher plants and is produced from methionine in essentially all tissues. Production of ethylene varies with the type of tissue, the plant species, and also the stage of development. The mechanism by which ethylene is produced from methionine is a 3 step process. ATP is an essential component in the synthesis of ethylene from methionine. ATP and water are added to methionine resulting in loss of the three phosphates and S-adenosyl methionine. 1-amino-cyclopropane-1-carboxylic acid synthase facilitates the production of ACC from SAM.

Oxygen is then needed in order ro oxidize ACC and produce ethylene. This reaction is catalyzed by an oxidative enzyme called ethylene forming enzyme. The control of ethylene production has received considerable study.

Study of ethylene has focused around the synthesis promoting effects of auxin, wounding, and drought as well as aspects of fruit-ripening. ACC synthase is the rate limiting step for ethylene production and it is this enzyme that is manipulated in biotechnology to delay fruit ripening in the"flavour saver" tomatoes.

Functions of Ethylene

Ethylene is known to affect the following plant processes:

- Stimulates the release of dormancy.
- Stimulates shoot and root growth and differentiation (triple response)
- May have a role in adventitious root formation.
- Stimulates leaf and fruit abscission.
- Stimulates Bromiliad flower induction.

- Induction of femaleness in dioecious flowers.
- Stimulates flower opening.
- Stimulates flower and leaf senescence.

GIBBERELLINS

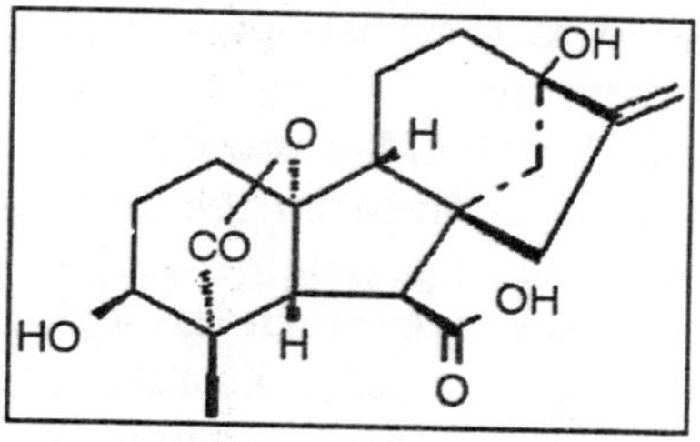

NATURE OF GIBBERELLINS

Unlike the classification of auxins which are classified on the basis of function, gibberellins are classified on the basis of structure as well as function. All gibberellins are derived from the ent-gibberellane skeleton. The structure of this skeleton derivative along with the structure of a few of the active gibberellins are shown above. The gibberellins are named GA1....GAn in order of discovery. Gibberellic acid, which was the first gibberellin to be structurally characterised, is GA3. There are currently 136 GAs identified from plants, fungi and bacteria

GA's are widespread and so far ubiquitous in both flowering (angiosperms) and non-flowering (gymnosperms) plants as well as ferns.

GIBBERELLIN BIOSYNTHESIS AND METABOLISM

Gibberellins are diterpenes synthesized from acetyl CoA via the mevalonic acid pathway. They all have either 19 or 20 carbon units grouped into either four or five ring systems. The fifth ring is a lactone ring as shown in the structures above attached to ring A. Gibberellins are believed to be synthesized in young tissues of the shoot and also the developing seed. It is uncertain whether young root tissues also produce gibberellins. There is also some evidence that leaves may be the source of some biosynthesis. The pathway by which gibberellins are formed is outlined below and illustrated in figure. 3 acetyl CoA molecules are oxidized by 2 NADPH molecules to produce 3 CoA molecules as a side product and mevalonic acid. Mevalonic acid is then Phosphorylated by ATP and decarboxylated to form isopentyl pyrophosphate. 4 of these molecules form geranylgeranyl pyrophosphate which serves as the donor for all GA carbon atoms.

This compound is then converted to copalylpyrophosphate which has 2 ring systems Copalylpyrophosphate is then converted to kaurene which has 4 ring systems.

Subsequent oxidations reveal kaurenol (alcohol form), kaurenal (aldehyde form), and kaurenoic acid respectively.

Kaurenoic acid is converted to the aldehyde form of GA12 by decarboxylation. GA12 is the 1st true gibberellane ring system with 20 carbons. From the aldehyde form of GA12 arise both 20 and 19 carbon gibberellins but there are many mechanisms by which these other compounds arise.

Certain commercial chemicals which are used to stunt growth do so in part because they block the synthesis of gibberellins. Some of these chemicals are Phosphon D, Amo-1618, Cycocel (CCC), ancymidol, and paclobutrazol. During active growth, the plant will metabolize most gibberellins by hydroxylation to inactive conjugates quickly with the exception of GA3. GA3 is degraded much slower which helps to explain why the symptoms initially associated with the hormone in the disease bakanae are present. Inactive conjugates might be stored or translocated via the phloem and xylem before their release (activation) at the proper time and in the proper tissue.

FUNCTIONS OF GIBBERELLINS

Active gibberellins show many physiological effects, each depending on the type of gibberellin present as well as the species of plant. Some of the physiological processes stimulated by gibberellins are outlined below:

- Stimulate stem elongation by stimulating cell division and elongation.

- Stimulates bolting/flowering in response to long days.
- Breaks seed dormancy in some plants which require stratification or light to induce germination.
- Stimulates enzyme production (a-amylase) in germinating cereal grains for mobilization of seed reserves.
- Induces maleness in dioecious flowers (sex expression).
- Can cause parthenocarpic (seedless) fruit development.
- Can delay senescence in leaves and citrus fruits.

GROWTH AND DEVELOPMENT

Growth is an increase in size of basic parts of plants (due to production of extra cells), whereas development is the production of new structures in the progress towards maturity - flowers, fruits etc.

PLANT HORMONES VS ANIMAL HORMONES

Both are

- Chemicals
- Produced in one region
- Move by diffusion not not blood flow to another ("target") region and have major physiological effects.

REGULATION OF PLANT GROWTH

In that green plants make growth in response to stimuli from their surroundings, they show sensitivity, i.e. one of the seven characteristic processes of living organisms. However, green plants respond to fewer stimuli than animals, and they make a slower and more restricted response.

The 3 stimuli to which green plants responds are:

- Light.
- Gravity.
- Water/touch (peasticks etc)NOT CO_2.

If a plant responds by growing in a particular direction,

it is called a tropism. Positive tropisms are towards the stimulus, negative tropisms are away from the stimulus.

REGULATION OF DEVELOPMENT

Stimulation of Roots in Cuttings

When cuttings are taken, the shoot which is selected is removed from the plant and disconnected from a supply of water and minerals from the roots. It is therefore important to encourage the development of roots, to aid the establishment of the"cutting" as an independent plant.

"Hormone rooting powders" contain the substance indoleacetic acid IAA which has been identified as the same chemical as"auxin" which causes curvature in roots and shoots. Usually the newly exposed surface of the cutting is dipped into a powder containing the hormone, mixed with inert powder, and then it is placed into a pot containing moist compost or soil, which is gently firmed around the cutting and placed in a cool moist environment.

There are a variety of other chemically similar compounds which may be used for this purpose.

HORMONE WEEDKILLERS

Although there are a wide variety of chemicals which can be toxic to plants or which damage their exposed parts, such as leaves and shoots, these are of limited use due to problems of toxicity to humans and other animals, and danger due to their oxidising nature. These substances may not affect parts of plants which are underground, and there is a likelihood that new growth will sprout again from these parts. There are also many situations in which more specificity is required: killing some plants but not others, and leaving soil fit for subsequent growth of plants which are required.

Hormone weedkillers have a variety of advantages over more traditional substances:

- They should be more toxic to plants than animals.
- They should be carried to underground plants and kill then too.
- Some weedkillers have a greater effect on"broad-

leaved weeds" than on grasses and also cereal crops (monocots). For this reason they are called selective weedkillers.

These may be used to control weeds amongst the grass on lawns, or to reduce competition from weeds in cereal crops. Examples include compounds with complicated chemical names, popularly abbreviated to 2.4-D and 2,4,5-T. Other compounds such as Dalapon are more effective against monocots than dicots, so they could be used in the opposite situation.

In fact these weedkilllers cause sensitive plants to grow too fast, damaging their cells so that their conducting tissue does not develop properly, or they use up their food reserves and die.

Weed

A plant growing in the wrong place - i.e. not planted by a gardener or grower. In other words, a plant which is able to compete and grow well in the prevailing conditions - probably in its own natural habitat!

Consequences of Using Selective Weedkillers

It must be admitted that the widespread use of selective weedkillers in European and Western agriculture has improved the efficiency of crop production (productivity of arable farms), in response to policies promoted by government ministries. This has been quite successful. resulting in the excessive production of cereal and other crops - so called"grain mountains".

However this apparent economic improvement has been at the expense of biodiversity in the countryside, as many"weed" species are food sources and act as shelter for many species of wildlife. Many plants which used to flower alongside crops and weeds, adding to the visual impact of the countryside, have been adversely affected by herbicides sprayed onto fields. Similarly, the survival of animal species, such as birds and butterflies, which add immeasurably to the public enjoyment of the wider environment, are consequently threatened more by the lack of food and protection than by the killing power of the chemicals involved.

Recently, there have been a series of developments aimed at reversing this trend, including"set-aside" where farmers are paid compensation in order to leave sections of their farms fallow in order to revert to"natural" conditions which should support normal fauna and flora. Regulations about maintenance of hedges and borders around fields have also been relaxed in order to increase the areas which can act as habitats for wildlife.

Development of Fruits

The ripening of fruits is known to be affected by a number of plant hormones. Commercially, growers sometimes use chemical sprays to prevent fruit from falling from trees. The substances used are auxins such as IAA or substances such as 2,4-D which is also used as a herbicide. Depending on weather conditions, this normally happens with a proportion of the fruit crop, leading to uncertainty as to the likely productivity of the crop. The consistency of cropping and ease of picking can thus be increased. Normally, a flower which is pollinated develops into a fruit, containing seeds derived from the fertilised egg cell. However, in some circumstances, fruit develops without pollination taking place. In seedless varieties of fruits this parthenocarpic fruit development occurs when other parts of the fruit produce auxins which would otherwise come from the seeds.

Other chemical substances acting as growth regulators can be used to induce flowering and alter the number and shape of fruit forming. The gas ethane (ethylene) has also been shown to have a hormone-like activity on fruit and is used to control the ripening of stored fruit such as bananas and apples which are best picked and shipped before maturity.

Chapter 6

Carbohydrate as Biomolecule

BIOMOLECULES

INTRODUCTION

This BioCoach activity will help you review the characteristics of the different classes of biomolecules: lipids, carbohydrates, proteins, and nucleic acids.

BUILDING BIOMOLECULES

This BioCoach activity will help you review molecular formulas and models, major elements and their valences, and the important functional groups of biological molecules. It will also let you try your hand at building certain molecular models.

GENE TO PROTEIN: TRANSCRIPTION

This BioCoach activity will help you review transcription, the synthesis of RNA from DNA. You will find information about the organization of genes, the structure of RNA, the transcription process, and the differences between mRNA in prokaryotes and eukaryotes. Animations show the transcription process in a dynamic way. You can test your understanding of transcription by using the Self-Quiz at the end of the module. By continuing on to the activity on translation, you can then learn about how mRNA is translated into protein.

For the Transcription and Translation activities, it may be helpful to review DNA structure and DNA replication.

GENE TO PROTEINS

This BioCoach activity will help you review translation, the synthesis of protein from mRNA. You will find information about the structure of proteins, the genetic code, the molecular components of translation, and the process of assembling a protein chain as a ribosome moves along an mRNA molecule. Animations show the translation process in a dynamic way. You can test your understanding of translation by using the Self-Quiz at the end of the module. It may be helpful to review other topics, such as DNA structure, DNA replication, transcription, and biomolecules.

BIOMOLECULE BIOGRAPHIES

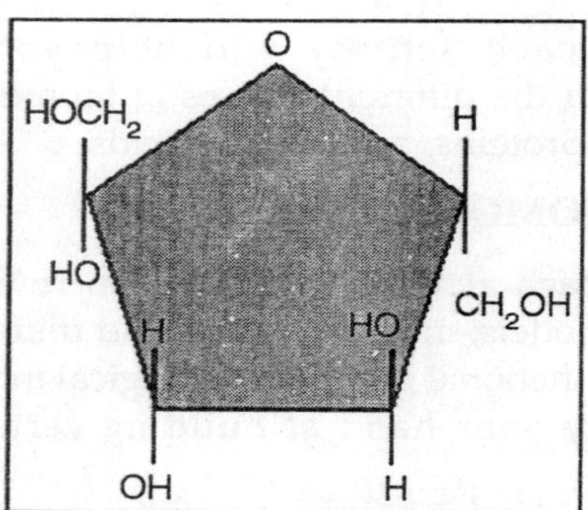

Fig. Fructose, a Monosacharide The Pentagon Structure Represents Five Carbon Atoms.

CARBOHYDRATES

The main function of carbohydrates in animals is to provide an energy source. A few carbohydrates serve as structural units such as deoxyribose. This sugar can be found in DNA. When the energy is not being used by the animal, the carbohydrate is stored as glycogen. Carbohydrates are composed of carbon, oxygen, and hydrogen. In most carbohydrates, there is one carbon atom for each water molecule. Carbohydrates are categorized into three groups based on molecule size.

MONOSACCHARIDES

Monosaccharides are composed of three to seven carbons. Prefixes indicate the number of carbon atoms contained in the molecule. For example, a pentose would consist of five carbon atoms. One of the main energy suppliers to the body happens to be a monosaccharide called glucose, a hexose.

DISACCHARIDES

A disaccharide is the result when two monosaccharides combine by dehydration synthesis. One water molecule also results as a byproduct. For example, the monosaccharides fructose and glucose combine in the chemical equation $C_6H_{12}O_6$(glucose) + $C_6H_{12}O_6$(fructose) $\rightarrow$ $C_{12}H_{22}O_{11}$(sucrose) + H_2O(water) to form sucrose and water. You may notice that glucose and fructose have the same chemical composition; however they differ in the relative positions of their carbon and oxygen atoms. In order to obtain monosaccharides from a disaccharide, all you need to do is add water. This is a process called hydrolysis.

POLYSACCHARIDES

Polysaccharides consist of not only three or four monosaccharides, but tens and hundreds in long chains. These monosaccharides were also joined by dehydration synthesis. The main polysaccharide that is in the human body is glycogen. Glycogen is stored in the liver and skeletal muscles and can be broken down into monosaccharides for use as energy. In order to achieve monosaccharide status, polysaccharides must also go through hydrolysis.

LIPIDS

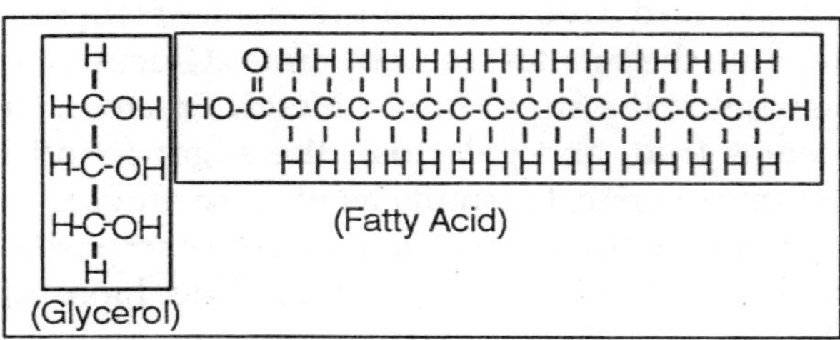

Lipids are very chemically diverse compunds which all share one common trait -- hydrophobia. They possess little or no affinity to water because of their molecular structure - mostly hydrocarbon regions with nonpolar bonds. There are three major categories of lipids:

FATS

large molecules composed of the two smaller molecules glycerol and fatty acids. When these two combine, the product is a triacylglycerol (or tryglyceride). The major function of fat is to supply a reserve of energy which is stored in adipose cells.

PHOSPHOLIPIDS

Similar to fats, but contain two fatty acids and a phosphate group which attaches to the glycerol molecule. The phosphate group is hydrophilic while the hydrocarbons remain hydrophobic. This structure allows phospholipids to act as cell membranes inside the body.

STEROIDS

Consist of four interconnected carbon rings. Cholesterol, an important steroid, is also a part of cell membranes.

BIOMOLECULES: THE CARBOHYDRATES

Carbohydrates are organic molecules with the general formula of CHO in 1:2:1 ratio. Although carbohydrates constitute only 1 to 2 Per cent of cell mass, they provide the raw fuel for cellular energy production. Carbohydrates are classified according to molecular size and solubility. In general, the smaller molecules are more soluble than the larger ones.

Monosaccharides include glucose, fructose, galactose, deoxyribose, and ribose.A single unit of sugar, monosaccharides are the smallest carbohydrates.Glucose, a 6-carbon sugar (hexose) is the sugar in our blood. Fructose, the sugar that sweetens fruit, and galactose, the sugar found in milk, have the same chemical formula as glucose and are therefore isomers of glucose. Isomers have the same composition but a different arrangement of their atoms and have different properties. Glucose can have a straight chain of carbon atoms or, more commonly, form a ring structure.Two other 5-carbon sugars or monosaccharides (called pentose) used in nucleic acid synthesis are deoxyribose and ribose.

Disaccharides include sucrose, lactose, and maltose.Two monosaccharides are joined together by dehydration synthesis to form a disaccharide molecule.Sucrose (table sugar)Glucose + fructose = sucrose + waterLactose (milk sugar)Glucose + galactose = lactose + waterMaltose (malt sugar)Glucose + glucose = maltose + water.

Polysaccharides include starch, cellulose, and glycogen.These long, chain-like polymers make ideal storage products due to their insolubility.Starch is the storage molecule synthesized from glucose by plants.Cellulose, which is also synthesized by plants for cell wall construction, is indigestible because we lack enzymes for it. Cellulose provides Fibre to promote peristalsis.Glycogen is the carbohydrate storage molecule found in muscle and liver cells. When blood sugar levels drop, liver cells hydrolyze glycogen and release glucose to the blood.

CLASSIFICATION OF CARBOHYDRATES

INTRODUCTION

The activities on which the life of the body depends involve a continuous expenditure of energy and a constant exchange of material. Ultimately the body is dependent upon food for the fuel materials which supply energy and for both the substances which are transformed in, and eliminated from, the body, and those whose presence regulates and controls these transformations.

The materials leaving the body are to be regarded not merely as wastes but as end products of an orderly and coordinated series of chemical reactions which occur in the body and by virtue of which its functions are performed.

Thus the chief functions of food are:

- To yield energy.
- To build tissue.
- To regulate body processes.

These functions involve reactions which are dependent upon the chemical composition and constitution of the food. Any food constituent which takes part in any of these functions may be regarded as having nutritive value.

Most of the nutrient material contained in food requires more or less change to bring it into the exact forms most useful in nutrition. These changes as a rule take place in the digestive tract and together constitute the process of digestion.

The changes which take place in the foodstuffs, after they have been absorbed from the digestive tract, are included under the general term"metabolism." Although the chemical changes and the energy transformations are of course inseparable, it has become customary to speak of the metabolism of matter and the metabolism of energy, and to regard the extent of the metabolism of any material substance as measured by the amount of its end products eliminated, and the extent of the energy metabolism as measured by the amount of heat, or of heat and external muscular work, which the body gives off.

The metabolism of matter and the metabolism of energy are normally supported by the food; but if no food is taken, they continue at the expense of the body substance. The expenditure of energy can never cease in the living body because it includes the work involved in carrying on the internal processes which are essential to life itself; and the expenditure of matter cannot cease because the energy for this necessary work is obtained by the breaking down of the organic compounds of the food or of the body substance into simpler compounds, many of which are of no further use to the body and must be eliminated. When the food supplies sufficient energy, the body substance is protected; when the food is insufficient, body substance is burned as fuel.

In order, then, to consider intelligently the nutritive requirements of the body as regards the substances of which it is composed, it is necessary first to know whether the fuel requirements (the requirements of the energy metabolism) have been fully met. The carbohydrates, fats, and proteins of the food all serve as fuel to yield the energy required for the activities of the body, and the proteins serve also as material for the maintenance or growth of body tissue. But of the fifteen chemical elements which are essential to the structure and functions of the body, simple proteins furnish only five. The remaining ten elements are largely constituents of the ash of the food and are known as ash constituents, inorganic foodstuffs, mineral matter, or salts.

Recent investigations have developed the fact that food of sufficient energy value and containing ample amounts of each of the chemical elements known to be essential to the body is not necessarily adequate to meet all the requirements of nutrition. Thus it appears that certain substances occurring in natural foods but not yet chemically identified are also to be included among the nutritive requirements of the body and therefore among the factors which determine the nutritive values of foods. At present these unidentified substances are referred to as"vitamines" or as"fat soluble A" and"water soluble B." *The essentials of a chemically adequate food supply may therefore be summarized as follows:*

- Sufficient of the organic nutrients in digestible forms to yield the needed energy.
- Protein, sufficient in amount and appropriate in kind.
- Adequate amounts and proper proportions of the various ash constituents or inorganic foodstuffs.
- Sufficient of each of the two unidentified"vitamine" factors, the"fat soluble A" and the"water soluble B."

In attempting to give in the following pages a general view of the chemistry of food and nutrition it has seemed best to discuss first the chemical nature and nutritive functions of the substances which serve as sources of energy in nutrition, then the nutritive requirements in terms of energy, protein, the more prominent"inorganic" elements, and the"vitamines," and finally the bearing of these various factors offood value upon problems connected with the economic use of food.

Of the constituents of the ordinary mixed food of man the carbohydrates are usually the most abundant and the most economical sources of energy. They are also considered to be the first of the three great groups of foodstuffs to be formed by synthesis from simple inorganic substances in plants;"in the long run, all the energy of living matter comes from them." The synthesis of carbohydrates in nature is therefore a logical starting point for the study of the organic foodstuffs.

In the chlorophyll cells of the leaves of green plants the energy of the sun's rays brings about reaction between carbon dioxide and water which results in the liberation of oxygen and the formation of organic compounds. There is still doubt as to the exact mechanism of the process and no certainty that it is the same in all cases. It has, however, been quite generally found that the volume of oxygen liberated is equal to that of carbon dioxide consumed. The simplest possible representation of the reaction would be $CO_2 + H_20 \rightarrow CH_20 + O_2$ according to which the first product of the synthesis would be formaldehyde. There is considerable (though not conclusive) evidence that formaldehyde is thus formed and that it is rapidly built up into less reactive compounds. Whatever the steps in the process* there is normally an early production of carbohydrate. Usually the first product which can be demonstrated as accumulating in the plant as the result of the photosynthesis is a sugar (glucose or sucrose) or starch. Assuming glucose as a typical product and neglecting the intermediate stages, the photosynthesis of carbohydrate may be represented thus:

$$6\ CO_2 + 6\ H_{20} ® C_6H_{12}O_6 + 6O_2$$

Glucose is the most familiar representative of a group of simple sugars (monosaccharides or monosaccharoses) which are in composition direct polymers of formaldehyde (CH20) and which are classified, according to the number of carbon atoms in the monosaccharide molecule, as trioses, pentoses, hexoses, etc.

CLASSIFICATION

Definitions of the term"simple sugar" vary somewhat, depending chiefly upon the views of different authors as to how simple a compound may properly be called a sugar.

According to Browne, a simple sugar or monosaccharide may be defined as an aldehyde alcohol or ketone alcohol of the aliphatic series, the molecule of which contains one carbonyl and one or more alcohol groups, one of the latter being always adjacent to the carbonyl group. All simple sugars contain, therefore, as a characteristic group upon the presence of which the chief chemical properties of the sugars depend.

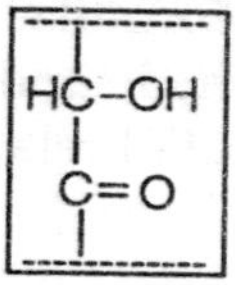

$$\begin{array}{c} | \\ HC-OH \\ | \\ C=O \\ | \end{array}$$

The simplest possible sugar according to this definition is glycolaldehyde, CH2OH-CHO, which (in analogy with the nomenclature of the familiar sugars) may also be called glycolose. The structural formulae of glucose and fructose, the most familiar representatives of the aldehyde-alcohol (aldose) and ketone-alcohol (ketose) sugars, respectively, are as follows:

Glucose

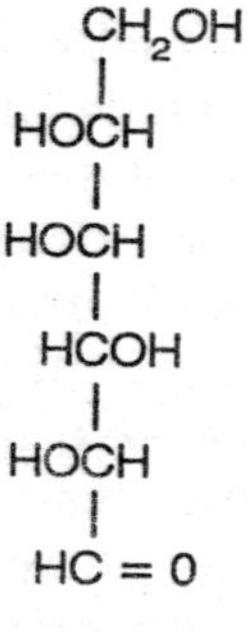

$$\begin{array}{c} CH_2OH \\ | \\ HOCH \\ | \\ HOCH \\ | \\ HCOH \\ | \\ HOCH \\ | \\ HC=0 \end{array}$$

Fructose

$$\begin{array}{c} CH_2OH \\ | \\ HOCH \\ | \\ HOCH \\ | \\ HCOH \\ | \\ C=0 \\ | \\ CH_2OH \end{array}$$

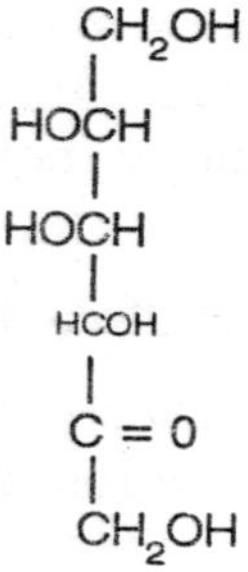

Since glucose gives aldehyde reactions but not so readily as the above structural formula would lead one to expect, it is believed that in ordinary solutions of glucose the substance exists partly in the condition indicated by the aldehyde formula and partly in a tautomeric form represented by the lactone or"oxygen bridge" formula.

Following are the aldehyde and lactone formulae written without reference to the spatial relationships of the hydrogen and hydroxyl groups:

Aldehyde form

$$CH_2OH\text{-}CHOH\text{-}CHOH\text{-}CHOH\text{-}CHOH\text{-}CHO$$

Lactone form

$$CH_2OH\text{—}CHOH\text{—}\overset{\ulcorner\text{———}O\text{———}\urcorner}{CH\text{—}CHOH\text{—}CHOH\text{—}CHOH}$$

The name monosaccharide ("single sugar") implies that the monosaccharide molecule contains only one sugar radicle - that it cannot be split by hydrolysis into sugars of lower molecular weight. A substance like cane sugar which on hydrolysis splits to two molecules of simple sugar is called adisaccharide or disaccha-rose ("double sugar"). Trisaccharides and tetrasaccharides are also known.

Substances which like starch are of high molecular weight and on complete hydrolysis yield many molecules of simple sugar are called polysaccharides * or polysaccharoses. The term"carbohydrates" covers all the simple sugars and all substances which can be converted into simple sugars by hydrolysis. The term"glucosides" is applied to substances which consist of combinations of carbohydrate radicles with radicles of other kinds and which therefore yield on hydrolysis both a simple sugar and one or more products of other than carbohydrate nature.

CARBOHYDRATES MONOSACCHARIDES

- (Monosaccharoses) Dioses ($C_2H_4O_2$) - Glycolose.
- Moses ($C_3H_6O_3$). Aldoses - Glycerose. Ketose - Dioxyacetone.

- Tetroses ($C_4H_8O_4$).
- Aldoses - Erythrose,3 Threose.3 Ketose - Erythrulose.2
- Pentoses ($C_6H_{10}O_5$).
- Aldoses - Arabinose,2 Xylose,2 Ribose,2 Lyxose.8 Ketoses
- - Araboketose3, Xyloketose (ketoxylose).2 [Methyl pentoses ($C_6H_{10}O_5$) - Rhamnose,2 Fucose2].

Some writers use the term polysaccharides to include all carbohydrates other than monosaccharides. Mathews applies it to all carbohydrates more complex than the disaccharides.

Names of a few of the most important carbohydrates are printed in small capitals. Separate mention of the d, l, and dl forms of the various sugars is omitted, since in the study of food and nutrition we are practically concerned only with that one of the three forms which is found in or derived from natural products.

- Hexoses ($C_6H_{12}O_6$).
- Aldoses - Glucose,1 Mannose,2 Galactose,2 Gulose,3 Idose,3Talose,3.
- Allose,2 Altrose.3 Ketoses - Fructose,1 Sorbose,2 Tagatose.3.
- Heptoses ($C_7H_{14}O_7$). Aldose - Mannoheptose.1 Ketose - Sedoheptose.1.

Note: No attempt is here made to summarize the occurrence of any but the tetroses, pentoses, hexoses and heptoses. Glycolose and the trioses if formed in nature are probably too reactive to accumulate sufficiently for identification.

Disaccharides (Disaccharoses)

Dihexoses (Hexobioses) - ($C_{12}H_{22}O_{11}$). Anhydride of glucose + fructose-Sucrose. Anhydrides of glucose + galactose - Lactose, Melibiose. Anhydrides of glucose + glucose- Maltose, Isomaltose, Trehalose, Turanose.

Trisaccharides (Trisaccharoses)

Trihezoses ($C_{18}H_{22}O_{16}$). Anhydride of glucose + galactose

+fructose - Raffinose. Anhydride of glucose + glucose + glucose - Melezitose. Anhydride of fructose + fructose + fructose - Secalose.

Tetrasaccharides (Tetrasaccharoses)

Tetrahexoses ($C_{24}H_{42}O_{21}$). Anhydrides of 2 galactose + glucose + fructose - Stachyose, Lupeose.

Polysaccharides (Polysaccharoses)

Pentosans (chief constituents of gums and mucilages). Anhydrides of xylose - Xylans. Anhydrides of arabinose - Arabans.

Hexosans

Anhydrides of glucose-Starch, Cellulose, Glycogen, Dextrin (and other"dextrans"). Anhydrides of mannose - Mannans. Anhydrides of galactose- Galactans (pectins). Anhydrides of fructose - Inulin (and other"levulans").

- Occurs free in nature.
- Not yet found free in nature (or only in small amounts) but obtained by hydrolysis or fermentation of natural product.
- Known only (with certainty) as a Labouratory product.

SIMPLE BIOCHEMICAL TESTS

Materials

- Benedict's solution (not"quantitative")
- Iodine solution (I/KI)
- Waterbaths set at about 100°C
- Boiling tubes Racks
- 1% glucose soln
- 1% fructose soln
- % sucrose soln
- other sugars?
- 1% starch sol solution X

HCl, NaOH, litmus/universal indicator paper filter

funnels, papers, flasks? measuring cylinders (10, 25, 50, 100 ml) plastic pipettes glass pipettes (1, 5, 10 ml?), pipette pumps

Objectives

Familiarisation with Benedict's test - for reducing sugars modification of Benedict's test for non-reducing sugars iodine test dilution techniques sensitivity of above tests.

Procedure

See also Roberts Students' manual p. 42 You will be expected to write up and hand in a concise account of your techniques and the results. There is no need to use large amounts of test substances or reagents.

SAFETY

Irritant Solutions Wear Safety Glasses And Lab Coats Hot Water And Steam Take Care: Keep Working Area Clear:

- Carry out Benedict's test on the named sugar solutions and starch as provided.
- Use fixed (noted!) quantities of test substance and reagents.
- Do not contaminate stock solutions shared by the group.
- Note results - colour and intensity - separately from conclusions.
  - *Refinement*: Filter the contents of the tube in which you have been testing glucose (using paper previously labelled with pencil). Consider the colour of the filtrate, and accordingly modify the quantities of reagents used. Repeat filtration process. Explain your conclusions in your write-up.
- Dilute the sugar solutions (say ×10, ×100 etc), and carry out Benedict's test on suitable samples (i.e. amounts same as before).
- Explain the dilution technique you used in your write-up.
- What does this tell you about the sensitivity of the test?

- Using sucrose, carry out the test for non reducing sugars. Having obtained a negative result from ordinary benedict's test, a second test is performed after hydrolysing it to reducing sugars (acid hydrolysis, followed by neutralisation), then performing Benedict's test:-
  Add... drops/ml HCl - leave for... minutes at room temperature (... °C) or... mins at 100 °C - add... drops/ml NaOH (same strength) - check neutrality with litmus (colour......) or universal indicator paper (colour.....).
- Test solution X using a combination of the above tests.
- Test starch solution using iodine solution (I/KI). You could use drops in a recessed tile, but for sensitivity purposes it is probably best to use boiling tubes and drops of iodine.
  - Dilute starch × 10, × 100, etc. and repeat (standardised) test.
  - Warm the last tubes - i.e. the least blue coloured (100 °C waterbath)), and observe colour.
  - Cool (the lower end of) the tube under the tap.
  - Conclusion?

Chapter 7

Metabolism and Glycolysis

CARBOHYDRATES METABOLISM

Carbohydrate metabolism begins with digestion in the small intestine where monosaccharides are absorbed into the blood stream.

Blood sugar concentrations are controlled by three hormones: insulin, glucagon, and epinephrine. If the concentration of glucose in the blood is too high, insulin is secreted by the pancreas. Insulin stimulates the transfer of glucose into the cells, especially in the liver and muscles, although other organs are also able to metabolize glucose.

In the liver and muscles, most of the glucose is changed into glycogen by the process of glycogenesis (anabolism). Glycogen is stored in the liver and muscles until needed at some later time when glucose levels are low.

If blood glucose levels are low, then eqinephrine and glucogon hormones are secreted to stimulate the conversion of glycogen to glucose. This process is called glycogenolysis(catabolism).

If glucose is needed immediately upon entering the cells to supply energy, it begins the metabolic process calledglycoysis (catabolism). The end products of glycolysis arepyruvic acid and ATP.

Since glycolysis releases relatively little ATP, further reactions continue to convert pyruvic acid to acetyl CoA and then citric acid in the citric acid cycle. The majority of the ATP is made from oxidations in the citric acid cycle in connection with the electron transport chain.

During strenuous muscular activity, pyruvic acid is converted into lactic acid rather thatn acetyl CoA. Durlng the resting period, the lactic acid is converted back to pyruvic acid. The pyruvic acid in turn is converted back to glucose by the process called gluconeogenesis(anabolism). If the glucose is not needed at that moment, it is converted into glycogen by glycogenesis. You can remember those terms if you think of"genesis" as the formation-beginning.

These processes are summarized in the Metaboism Summary in the graphic on the left.. Each of these processes will be developed in greater detail various pages of this module.

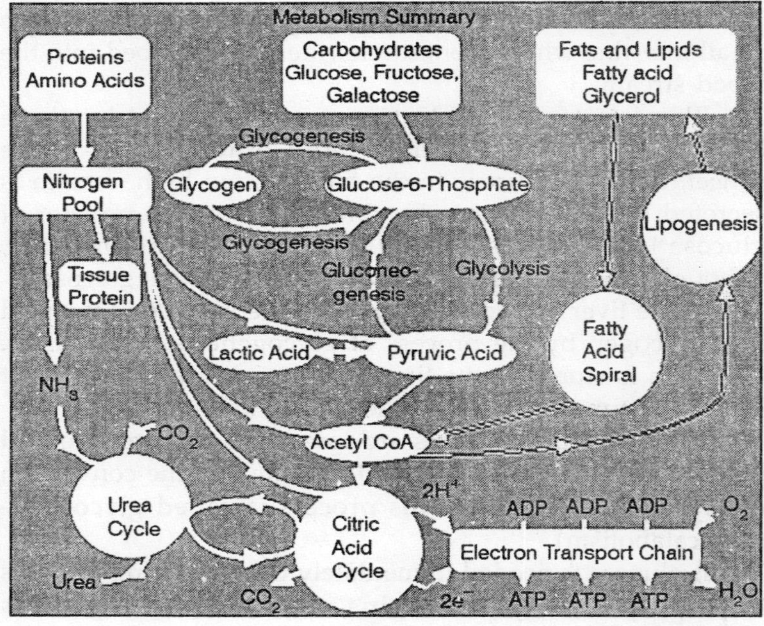

GLYCOLYSIS

All organisms produce ATP by releasing energy stored in glucose and other sugars.

- Plants make ATP during photosynthesis.
- All other organisms, including plants, must produce ATP by breaking down molecules such as glucose.

AEROBIC RESPIRATION

The process by which a cell uses O2 to"burn" molecules and release energy.

- *The reaction*: $C_6H_{12}O_6 + 6O_2 >> 6CO_2 + 6H_2O$.
- This reaction is the opposite of photosynthesis

This reaction takes place over the course of three major reaction pathways.

- Glycolysis.
- The Krebs Cycle.
- Electron Transport Phosphorylation (chemiosmosis) Glycolysis (glyco = sugar; lysis = breaking).
- *Goal*: Break glucose down to form two pyruvates.
- *Who*: All life on earth performs glyclolysis
- *Where*: The cytoplasm.
- Glycolysis produces 4 ATP's and 2 NADH, but uses 2 ATP's in the process for a net of 2 ATP and 2 NADH.

This process does not require O_2 and does not yield much energy.

FIRST STAGE OF GLYCOLYSIS

- Glucose (6C) is broken down into 2 PGAL's (Phosphoglyceraldehyde - 3Carbon molecules).
- This requires two ATP's.

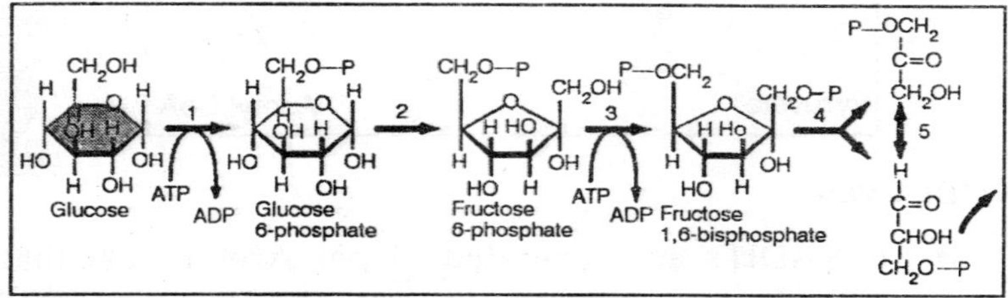

SECOND STAGE OF GLYCOLYSIS

- 2 PGAL's (3C) are converted to 2 pyruvates.
- This creates 4 ATP's and 2 NADH's.

- The net ATP production of Glycolysis is 2 ATP's.

- Goal: take pyruvate and put it into the Krebs cycle, producing NADH and $FADH_2$.
- *Where*: The mitochondria.
- There are two steps.
  - The Conversion of Pyruvate to Acetyl CoA.
  - The Krebs Cycle proper.
- The Krebs cycle and the conversion of pyruvate to Acetyl CoA produce 2 ATP's, 8 NADH's, and 2FADH2's per glucose molecule.

Oxidation of Pyruvate to form Acetyl CoA for Entry Into the Krebs Cycle.

- 2 NADH's are generated (1 per pyruvate).
- 2 CO_2 are released (1 per pyruvate).

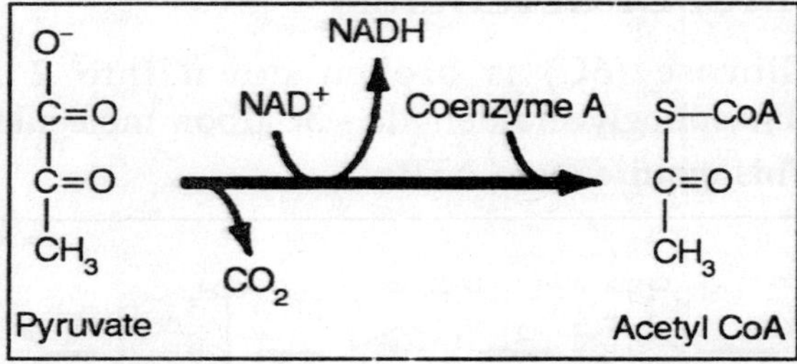

Krebs Cycle

- 6 NADH's are generated (3 per Acetyl CoA that enters).
- 2 $FADH_2$ is generated (1 per Acetyl CoA that enters).
- 2 ATP are generated (1 per Acetyl CoA that enters).
- 4 CO_2's are released (2 per Acetyl CoA that enters).

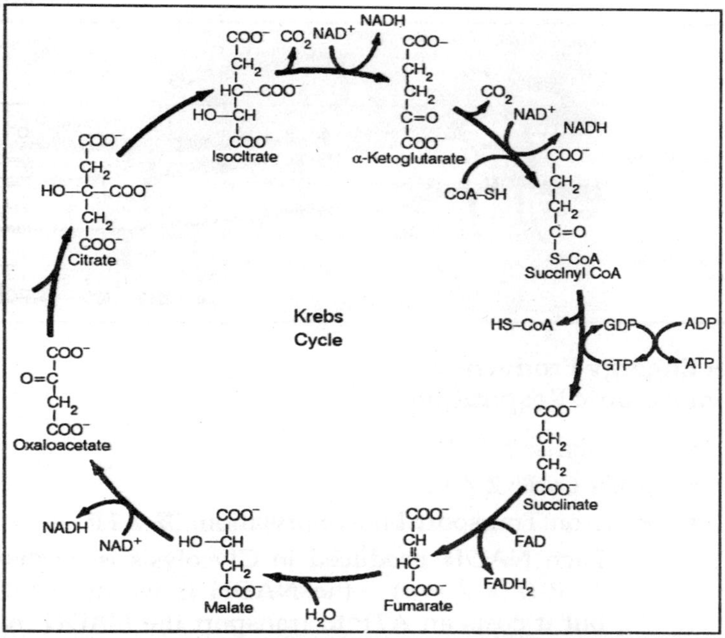

- Therefore, the total numbers of molecules generated in the oxidation of pyruvate and the Krebs Cycle is:
  - 8 NADH
  - 2 $FADH_2$
  - 2 ATP
  - 6 CO_2

Electron Transport Phosphorylation (Chemiosmosis)

- *Goal*: to break down NADH and $FADH_2$, pumping H+ into the outer compartment of the mitochondria
- *Where*: the mitochondria.
- In this reaction, the ETS creates a gradient which is used to produce ATP, quite like in the chloroplast
- Electron Transport Phosphorylation typically produces 32 ATP's.
- ATP is generated as H+ moves down its concentration gradient through a special enzyme called ATP synthase.

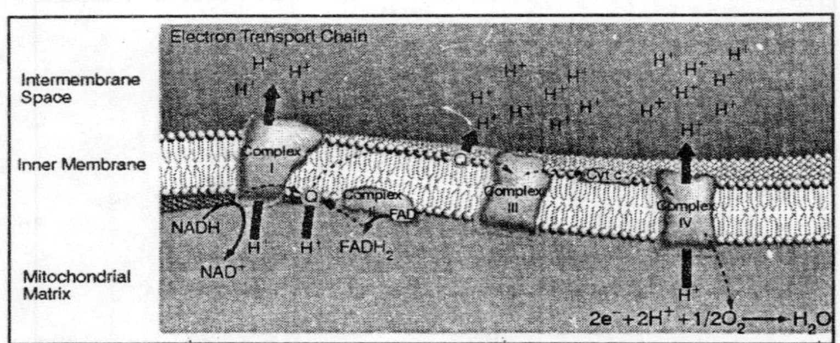

Net Engergy Production from Aerobic Respiration

- *Glycolysis*: 2 ATP.
- *Krebs Cycle*: 2 ATP.
- Electron Transport Phosphorylation: 32 ATP.
  - Each NADH produced in Glycolysis is worth 2 ATP (2 × 2 = 4) – the NADH is worth 3 ATP, but it costs an ATP to transport the NADH into the mitochondria, so there is a net gain of 2 ATP for each NADH produced in gylcolysis.
  - Each NADH produced in the conversion of pyruvate to acetyl COA and Krebs Cycle is worth 3 ATP (8 × 3 = 24).
  - Each FADH2 is worth 2 ATP (2 × 2 = 4).
  - 4 + 24 + 4 = 32.
- Net Energy Production: 36 ATP!

Anaerobic Respiration

- *Goal*: to reduce pyruvate, thus generating NAD+.
- *Where*: the cytoplasm.
- *Why*: in the absence of oxygen, it is the only way to generate NAD+.
- Alcohol Fermentation - occurs in yeasts in many bacteria.
  - The product of fermentation, alcohol, is toxic to the organism.

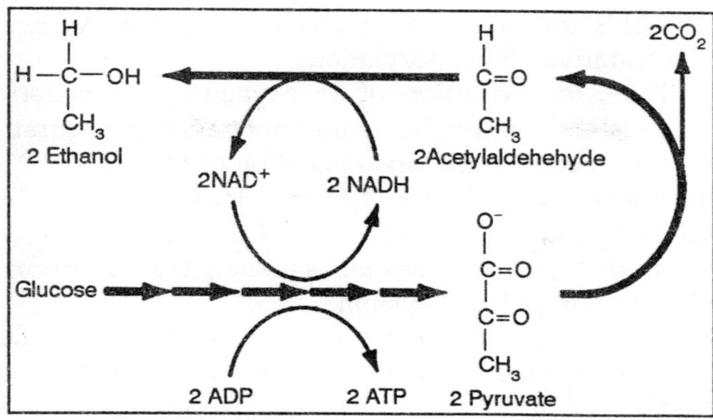

- *Lactic Acid Fermentation*: Occurs in humans and other mammals.
  - The product of Lactic Acid fermentation, lactic acid, is toxic to mammals.
  - This is the"burn" felt when undergoing strenuous activity.

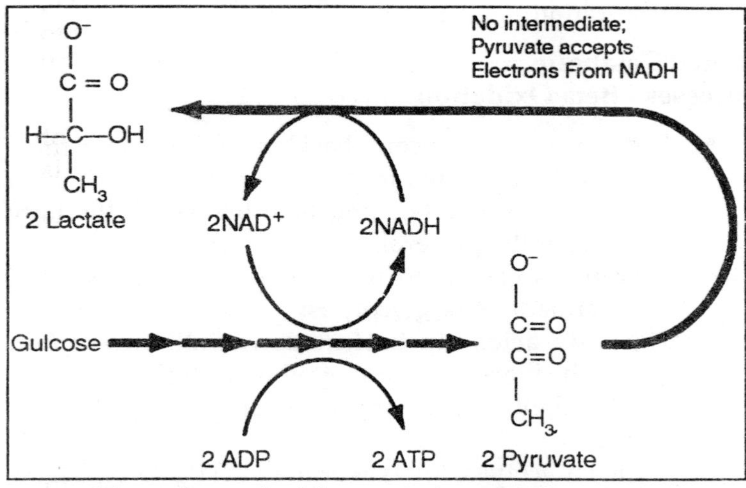

- The only goal of fermentation reactions is to convert NADH to NAD+ (to use in glycolysis).
- No energy is gained.
- Note differences - Anaerobic respiration - 2 ATP's produced (from glycolysis), aerobic respiration - 36

ATP's produced (from glycolysis, Krebs cycle, and Oxidative Phosphorylation).

- Thus, the evolution of an oxygen-rich atmosphere, which facilitated the evolution of aerobic respiration, was crucial in the diversification of life.

Photosynthesis: 6 CO_2 + 6 H_2O >> $C_6H_{12}O_6$ + 6 O_2.
Respiration: $C_6H_{12}O_6$ + 6 O_2 >> 6 CO_2 + 6 H_2O.

We see that these reactions are opposites: This is important since the earth is a closed system.

All life has a set amount of natural materials to work with, so it is important that they all be cycled through effectively and evenly.

Energy Yields

- *Glucose*: 686 kcal/mol.
- *ATP*: 7.5 kcal/mol.
- 7.5 × 36 = 270 kcal/mol for all ATP's produced.
- 270/ 686 = 39% energy recovered from aerobic respiration

Related Catabolic Processes - Beta Oxidation

- Fats consist of a glycerol backbone with two or three fatty acids connected to it.
- The body absorbs fats and then breaks off the fatty acids from the glycerol.
- Glycerol is converted to glyceraldehyde phosphate, an intermediate of glycolysis.
- The fatty acids are broken down into two-carbon units which are then converted to acetyl CoA.
  - An eight-carbon fatty acid can produce 4 acetyl CoA's.
  - Each acetyl CoA is worth 12 ATP's (3 NADP, 1 $FADH_2$, 1 ATP).
  - Therefore, this short fatty acid is worth 48 ATP's, a fat with three chains of this length would be worth 144 ATP's!
  - This is why fats are such a good source of

energy, and are hard to lose if you want to lose weight.

Comparison Between Plants and Animals

- Animal cells and Plant cells contain mitochondria!
  - However, animal cells contain many more mitochondria than plant cells.
- Animal cells get most of their ATP from mitochondria.
- Plant cells get most of their ATP from the chloroplast
  - The ATP generated from the mitochondria is only used when the plant cannot generate ATP directly from the light-dependent reactions.

Molecules used in Glycolysis and Krebs Cycle

- Not all of the molecules that enter Glycolysis and the Krebs Cycle are used for energy.
- Some are used to synthesize fats, nucleotides, amino acids, and other biologically important molecules.

Regulation of Glycolysis and the Krebs Cycle

- *Step 3 of Glycolysis*: The conversion of Fructose 6-phosphate to Fructose 1,6-bisphosphate.
  - Enzyme catalyzing this reaction = Phosphofructokinase.
  - *"Committing Step"*: Fructose 6-phosphate can be used by the cell for lots of things, but fructose 1,6-bisphosphate has limited use except in glycolysis.
  - Phosphofructokinase inhibited by high levels of ATP.
- ATP is also a substrate - odd, eh?
- Enzyme has two ATP binding sites, one in the active site and one in an allosteric site.
- Low to mid levels of ATP cause ATP to bind to the active site.
- High levels of ATP also enable ATP to bind to allosteric site, causing a conformation change and shutting down the enzyme.

- Conversion of Pyruvate to Acetyl CoA.
  - Enzyme involved in catalyzing this reaction = pyruvate dehydrogenase.
  - High levels of ATP slow down this reaction by phosporylating the enzyme, changing its shape and shutting it down.
- High levels of NADH and Acetyl CoA also inhibit this enzyme.
  - NAD+, CoA, or AMP (an indicator of low ATP) can speed up the reaction.

Chapter 8

TCA Cycle

OVERVIEW

The bulk of ATP used by many cells to maintain homeostasis is produced by the oxidation of pyruvate in the TCA cycle. During this oxidation process, reduced nicotinamide adenine dinucleotide (NADH) and reduced flavin adenine dinucleotide (FADH2) are generated. The NADH and $FADH_2$ are principally used to drive the processes of oxidative phosphorylation, which are responsible for converting the reducing potential of NADH and $FADH_2$ to the high energy phosphate in ATP.

The fate of pyruvate depends on the cell energy charge. In cells or tissues with a high energy charge pyruvate is directed towards gluconeogenesis, but when the energy charge is low pyruvate is preferentially oxidized to CO_2 and H_2O in the TCA cycle, with generation of 15 equivalents of ATP per pyruvate. The enzymatic activities of the TCA cycle (and of oxidative phosphorylation) are located in the mitochondrion. When transported into the mitochondrion, pyruvate encounters two principal metabolizing enzymes: pyruvate carboxylase (a gluconeogenic enzyme) and pyruvate dehydrogenase (PDH), the first enzyme of the PDH complex.

With a high cell-energy charge coenzyme A (CoA) is highly acylated, principally as acetyl-CoA, and able allosterically to activate pyruvate carboxylase, directing pyruvate towards gluconeogenesis. When the energy charge is low CoA is not acylated, pyruvate carboxylase is inactive, and pyruvate is preferentially metabolized via the PDH complex and the enzymes of the TCA cycle to CO2and H2O. Reduced NADH and FADH2 generated during the oxidative reactions can then be used to drive ATP synthesis via oxidative phosphorylation. The PDH complex is comprised of multiple copies of 3 separate enzymes: pyruvate dehydrogenase (20-30 copies), dihydrolipoyl transacetylase (60 copies) and dihydrolipoyl dehydrogenase (6 copies).

The complex also requires 5 different coenzymes: CoA, NAD+, FAD+, lipoic acid and thiamine pyrophosphate (TPP). Three of the coenzymes of the complex are tightly bound to enzymes of the complex (TPP, lipoic acid and FAD+) and two are employed as carriers of the products of PDH complex activity (CoA and NAD+). The pathway of PDH oxidation of pyruvate to acetyl-CoA is diagrammed below.

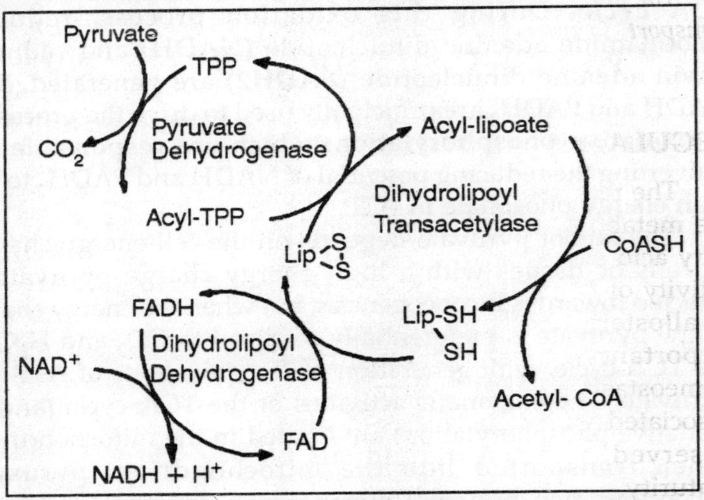

Flow diagram depicting the overall activity of the pyruvate dehydrogenase complex. During the oxidation of pyruvate to CO_2 by pyruvate dehydrogenase the electrons flow from pyruvate to the lipoamide moiety of dihydrolipoyl transacetylase then to the FAD cofactor of dihydrolipoyl dehydrogenase and finally to reduction of NAD+ to NADH. The acetyl group is linked to coenzyme A (CoASH) in a high energy thioester bond. The acetyl-CoA then enters the TCA cycle for complete oxidation to CO_2 and H_2O.

The first enzyme of the complex is PDH itself which oxidatively decarboxylates pyruvate. During the course of the reaction the acetyl group derived from decarboxylation of pyruvate is bound to TPP. The next reaction of the complex is the transfer of the two carbon acetyl group from acetyl-TPP to lipoic acid, the covalently bound coenzyme of lipoyl transacetylase. The transfer of the acetyl group from acyl-lipoamide to CoA results in the formation of 2 sulfhydryl (SH) groups in lipoate requiring reoxidation to the disulfide (S-S) form to regenerate lipoate as a competent acyl acceptor. The enzyme dihydrolipoyl dehydrogenase, with FAD+ as a cofactor, catalyzes that oxidation reaction. The final activity of the PDH complex is the transfer of reducing equivalents from the FADH2 of dihydrolipoyl dehydrogenase to NAD+.

The fate of the NADH is oxidation via mitochondrial electron transport, to produce 3 equivalents of ATP:

$$\text{Pyruvate} + \text{CoA} + \text{NAD+} \rightarrow CO_2 + \text{acetyl-CoA} + \text{NADH} + \text{H+}$$

REGULATION OF PDH COMPLEX

The reactions of the PDH complex serves to interconnect the metabolic pathways of glycolysis, gluconeogenesis and fatty acid synthesis to the TCA cycle. As a consequence, the activity of the PDH complex is highly regulated by a variety of allosteric effectors and by covalent modification. The importance of the PDH complex to the maintenance of homeostasis is evident from the fact that although diseases associated with deficiencies of the PDH complex have been observed, affected individuals often do not survive to maturity.

Since the energy metabolism of highly aerobic tissues such as the brain is dependent on normal conversion of pyruvate to acetyl-CoA, aerobic tissues are most sensitive to deficiencies in components of the PDH complex. Most genetic diseases associated with PDH complex deficiency are due to mutations in PDH. The main pathologic result of such mutations is moderate to severe cerebral lactic acidosis and encephalopathies.

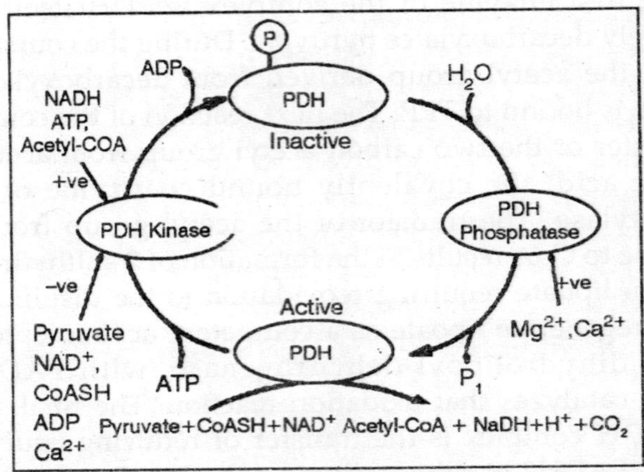

Factors regulating the activity of pyruvate dehydrogenase, (PDH). PDH activity is regulated by its' state of phosphorylation, being most active in the dephosphorylated state. Phosphorylation of PDH is catalyzed by a specific PDH kinase. The activity of the kinase is enhanced when cellular energy charge is high which is reflected by an increase in the level of ATP, NADH and acetyl-CoA. Conversely, an increase in pyruvate strongly inhibits PDH kinase. Additional negative effectors of PDH kinase are ADP, NAD+ and CoASH, the levels of which increase when energy levels fall. The regulation of PDH phosphatase is not completely understood but it is known that Mg2+and Ca2+ activate the enzyme. In adipose tissue insulin increases PDH activity and in cardiac muscle PDH activity is increased by catecholamines.

Two products of the complex, NADH and acetyl-CoA, are negative allosteric effectors on PDH-a, the non-phosphorylated, active form of PDH. These effectors reduce the affinity of the enzyme for pyruvate, thus limiting the flow of carbon through the PDH complex. In addition, NADH and acetyl-CoA are powerful positive effectors on PDH kinase, the enzyme that inactivates PDH by converting it to the phosphorylated PDH-b form. Since NADH and acetyl-CoA accumulate when the cell energy charge is high, it is not surprising that high ATP levels also up-regulate PDH kinase activity, reinforcing down-regulation of PDH activity in energy-rich cells. Note, however, that pyruvate is a potent negative effector on PDH kinase, with the result that when pyruvate levels rise, PDH-a will be favoured even with high levels of NADH and acetyl-CoA.

Concentrations of pyruvate which maintain PDH in the active form (PDH-a) are sufficiently high so that, in energy-rich cells, the allosterically down-regulated, high Km form of PDH is nonetheless capable of converting pyruvate to acetyl-CoA. With large amounts of pyruvate in cells having high energy charge and high NADH, pyruvate carbon will be directed to the 2 main storage forms of carbon (glycogen via gluconeogenesis and fat production via fatty acid synthesis) where acetyl-CoA is the principal carbon donor.

Although the regulation of PDH-b phosphatase is not well understood, it is quite likely regulated to maximize pyruvate oxidation under energy-poor conditions and to minimize PDH activity under energy-rich conditions.

REACTIONS OF THE TCA CYCLE

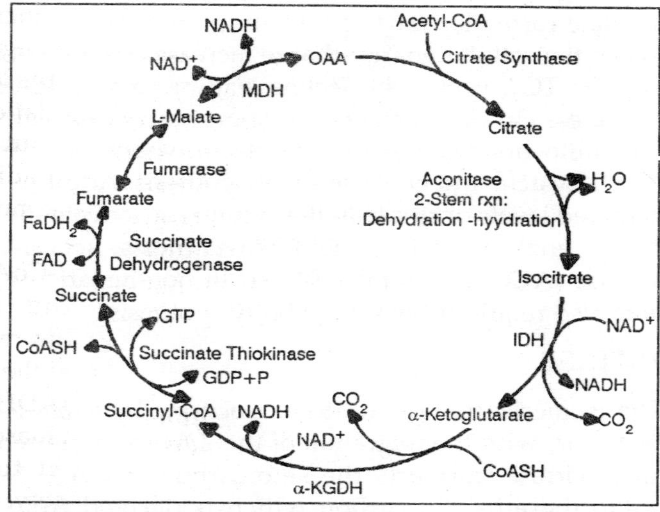

The TCA cycle showing enzymes, substrates and products. The GTP generated during the succinate thiokinase (succinyl-CoA synthetase) reaction is equivalent to a mole of

ATP by virtue of the presence of nucleoside diphosphokinase. The 3 moles of NADH and 1 mole of $FADH_2$ generated during each round of the cycle feed into the oxidative phosphorylation pathway. Each mole of NADH leads to 3 moles of ATP and each mole of $FADH_2$leads to 2 moles of ATP. Therefore, for each mole of pyruvate which enters the TCA cycle, 12 moles of ATP can be generated. IDH = isocitrate dehydrogenase. α-KGDH = α-ketoglutarate dehydrogenase. MDH = malate dehydrogenase.

CITRATE SYNTHASE (CONDENSING ENZYME)

The first reaction of the cycle is condensation of the methyl carbon of acetyl-CoA with the keto carbon (C-2) of oxaloacetate (OAA). The standard free energy of the reaction, -8.0 kcal/mol, drives it strongly in the forward direction. Since the formation of OAA from its precursor is thermodynamically unfavourable, the highly exergonic nature of the citrate synthase reaction is of central importance in keeping the entire cycle going in the forward direction, since it drives oxaloacetate formation by mass action principals.

When the cellular energy charge increases the rate of flux through the TCA cycle will decline leading to a build-up of citrate. Excess citrate is used to transport acetyl-CoA carbons from the mitochondrion to the cytoplasm where they can be used for fatty acid and cholesterol biosynthesis. Additionally, the increased levels of citrate in the cytoplasm activate the key regulatory enzyme of fatty acid biosynthesis, acetyl-CoA carboxylase (ACC) and inhibit PFK-1. In non-hepatic tissues citrate is also required for ketone body synthesis.

ACONITASE

The isomerization of citrate to isocitrate by aconitase is stereospecific, with the migration of the -OH from the central carbon of citrate (formerly the keto carbon of OAA) being always to the adjacent carbon which is derived from the methylene ($-CH_2-$) of OAA. The stereospecific nature of the isomerization determines that the CO_2 lost, as isocitrate is oxidized to succinyl-CoA, is derived from the oxaloacetate used in citrate synthesis.

Aconitase is one of several mitochondrial enzymes known as non-heme-iron proteins. These proteins contain inorganic iron and sulfur, known as iron sulfur Centres, in a coordination complex with cysteine sulfurs of the protein. There are two prominent classes of non-heme-iron complexes, those containing two equivalents each of inorganic iron and sulfur Fe_2S_2, and those containing 4 equivalents of each Fe_4S_4. Aconitase is a member of the Fe_4S_4 class. Its iron sulfur Centres are often designated as $Fe_4S_4Cys_4$, indicating that 4 cystine sulfur atoms are involved in tghe complete structure of the complex. In iron sulfur compounds the iron is generally involved in oxidation-reduction events.

ISOCITRATE DEHYDROGENASE

Isocitrate is oxidatively decarboxylated to α-ketoglutarate by isocitrate dehydrogenase, (IDH). There are two different IDH enzymes. The IDH of the TCA cycle uses NAD^+ as a cofactor, whereas the other IDH uses $NADP^+$ as a cofactor. Unlike the NAD+-requiring enzyme, which is located only in the mitochondrial matrix, the $NADP^+$-requiring enzyme is found in both the mitochondrial matrix and the cytosol. IDH catalyzes the rate-limiting step, as well as the first NADH-yielding reaction of the TCA cycle. The CO_2 produced by the IDH reaction is the original C-1 carbon of the oxaloacetate used in the citrate synthase reaction.

It is generally considered that control of carbon flow through the cycle is regulated at IDH by the powerful negative allosteric effectors NADH and ATP and by the potent positive effectors; isocitrate, ADP and AMP. From the latter it is clear that cell energy charge is a key factor in regulating carbon flow through the TCA cycle.

α -KETOGLUTARATE DEHYDROGENASE COMPLEX

α-ketoglutarate is oxidatively decarboxylated to succinyl-CoA by the α-ketoglutarate dehydrogenase (α-KGDH) complex. This reaction generates the second TCA cycle equivalent of CO_2 and NADH. This multienzyme complex is very similar to the PDH complex in the intricacy of its protein makeup, cofactors, and its mechanism of action. Also, as with the PDH complex, the reactions of the α-KGDH complex proceed with a large negative standard free energy change. Although the α-KGDH of the complex is not subject to covalent modification, allosteric regulation is quite complex, with activity being regulated by energy charge, the NAD^+/NADH ratio, and effector activity of substrates and products..

Succinyl-CoA and α-ketoglutarate are also important metabolites outside the TCA cycle. In particular, α-ketoglutarate represents a key anapleurotic metabolite linking the entry and exit of carbon atoms from the TCA cycle to pathways involved in amino acid metabolism. α-ketoglutarate is also important for driving the malate-aspartate shuttle.

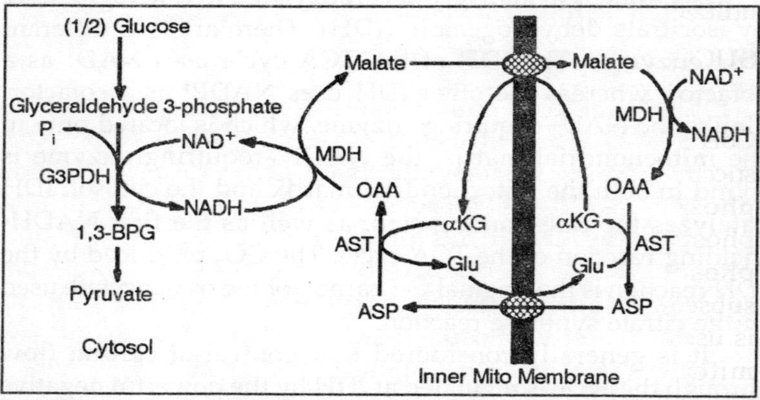

The malate/aspartate shuttle is the principal mechanism for the movement of reducing equivalents (in the form of NADH) from the cytoplasm to the mitochondria. The glycolytic pathway is a primary source of NADH. Within the mitochondria the electrons of NADH can be coupled to ATP production during the process of oxidative phosphorylation. The electrons are"carried" into the mitochondria in the form of malate. Cytoplasmic malate dehydrogenase (MDH) reduces oxaloacetate (OAA) to malate while oxidizing NADH to NAD+. Malate then enters the mitochondria where the reverse reaction is carried out by mitochondrial MDH. Movement of mitochondrial OAA to the cytoplasm to maintain this cycle requires it be transaminated to aspartate (Asp, D) with the amino group being donated by glutamate (Glu, E). The Asp then leaves the mitochondria and enters the cytoplasm. The deamination of glutamate generates α-ketoglutarate (α-KG) which leaves the mitochondria for the cytoplasm. All the participants in the cycle are present in the proper cellular compartment for the shuttle to function due to concentration dependent movement. When the energy level of the cell rises the rate of mitochondrial oxidation of NADH to NAD^+ declines and therefore, the shuttle slows. G3PDH is glyceraldehyde-3-phosphate dehydrogenase.

Succinyl-CoA, along with glycine, contributes all the carbon and nitrogen atoms required for the synthesis ofprotoporphyrin heme biosynthesis and for non-hepatic tissue utilization of ketone bodies.

SUCCINYL COA SYNTHETASE

The conversion of succinyl-CoA to succinate by succinyl CoA synthetase involves use of the high-energy thioester of succinyl-CoA to drive synthesis of a high-energy nucleotide phosphate, by a process known assubstrate-level phosphorylation. In this process a high energy enzyme--phosphate intermediate is formed, with the phosphate subsequently being transferred to GDP. Mitochondrial GTP is used in a trans-phosphorylation reaction catalyzed by the mitochondrial enzyme nucleoside diphospho kinase to phosphorylate ADP, producing ATP and regenerating GDP for the continued operation of succinyl CoA synthetase.

SUCCINATE DEHYDROGENASE (SDH)

Succinate dehydrogenase catalyzes the oxidation of succinate to fumarate with the sequential reduction of enzyme-bound FAD and non-heme-iron. In mammalian cells the final electron acceptor is coenzyme Q10 (CoQ10), a mobile carrier of reducing equivalents that is restricted by its lipophilic nature to the lipid phase of the mitochondrial membrane.

FUMARASE (FUMARATE HYDRATASE)

The fumarase-catalyzed reactions specific for the trans form of fumarate. The result is that the hydration of fumarate proceeds stereospecifically with the production of L-malate.

MALATE DEHYDROGENASE (MDH)

L-malate is the specific substrate for MDH, the final enzyme of the TCA cycle. The forward reaction of the cycle, the oxidation of malate yields oxaloacetate (OAA). In the forward direction the reaction has a standard free energy of about +7 kcal/mol, indicating the very unfavourable nature of the forward direction. As noted earlier, the citrate synthase reaction that condenses oxaloacetate with acetyl-CoA has a standard free energy of about -8 kcal/mol and is responsible for pulling the MDH reaction in the forward direction. The overall change in standard free energy change is about -1 kcal/mol for the conversion of malate to oxaloacetate and on to succinate.

The overall stoichiometry of the TCA cycle is:

acetyl-CoA + 3NAD$^+$ + FAD + GDP + P_i + $2H_2O \rightarrow 2CO_2$ + 3NADH + $FADH_2$ + GTP + 2H$^+$ + HSCoA

REGULATION OF THE TCA CYCLE

Regulation of the TCA cycle. like that of glycolysis, occurs at both the level of entry of substrates into the cycle as well as at the key reactions of the cycle. Fuel enters the TCA cycle primarily as acetyl-CoA. The generation of acetyl-CoA from carbohydrates is, therefore, a major control point of the cycle. This is the reaction catalyzed by the PDH complex.

By way of review, the PDH complex is inhibited by acetyl-CoA and NADH and activated by non-acetylated CoA (CoASH) and NAD$^+$. The pyruvate dehydrogenase activities of the PDH complex are regulated by their state of phosphorylation. This modification is carried out by a specific kinase (PDH kinase) and the phosphates are removed by a specific phosphatase (PDH phosphatase). The phosphorylation of PDH inhibits its activity and, therefore, leads to decreased oxidation of pyruvate. PDH kinase is activated by NADH and acetyl-CoA and inhibited by pyruvate, ADP, CoASH, Ca^{2+} and Mg^{2+}. The PDH phosphatase, in contrast, is activated by Mg^{2+} and Ca^{2+}.

Since three reactions of the TCA cycle as well as PDH utilize NAD+ as co-factor it is not difficult to understand why the cellular ratio of NAD+/NADH has a major impact on the flux of carbon through the TCA cycle.

Substrate availability can also regulate TCA flux. This occurs at the citrate synthase reaction as a result of reduced availability of oxaloacetate. Product inhibition also controls the TCA flux, e.g. citrate inhibits citrate synthase, α-KGDH is inhibited by NADH and succinyl-CoA. The key enzymes of the TCA cycle are also regulated allosterically by Ca^{2+}, ATP and ADP.

Chapter 9

Gluconeogenesis

INTRODUCTION

Gluconeogenesis occurs mainly in liver. Gluconeogenesis occurs to a more limited extent in the kidney and small intestine under some conditions.

Synthesis of glucose from pyruvate utilizes many of the same enzymes as Glycolysis. Three reactions of Glycolysis have such a large negative G in the forward direction that they are essentially irreversible (see lecture notes on Glycolysis):

HEXOKINASE, HOSPHOFRUCTOKINASE AND PYRUVATE KINASE

These steps must be bypassed in Gluconeogenesis. Two of the bypass reactions involve simple hydrolysis reactions.

Below is the forward reaction catalyzed by each of these Glycolysis enzymes, followed by the bypass reaction catalyzed by the Gluconeogenesis enzyme.

Hexokinase or Glucokinase Catalyzes

- Glucose + ATP' glucose-6-phosphate + ADP
- *Glucose-6-phosphatase (Gluconeogenesis) catalyzes*:
- Glucose-6-phosphate + H_2O' glucose + Pi

Glucose-6-phosphatase enzyme is embedded in the endoplasmic reticulum (ER) membrane in liver cells. Evidence indicates that the catalytic site is exposed to the ER lumen. Another subunit of the enzyme is postulated to function as a translocase, providing access of substrate to the active site.

Glucose-6-phosphatase

Glucose-6-phosphate $\xrightarrow{H_2O}$ Glucose $+ P_i$

Phosphofructokinase (Glycolysis) catalyzes: Fructose-6-phosphate + ATP' fructose-1,6-bisphosphate + ADP

- Fructose-1,6-bisphosphatase (Gluconeogenesis) catalyzes:
- Fructose-1,6-bisphosphate + H2O' fructose-6-phosphate + Pi

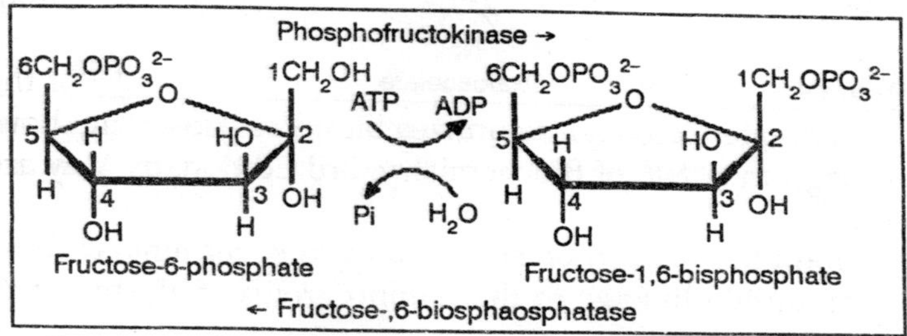

Pyruvate Kinase (last step of Glycolysis) catalyzes: phosphoenolpyruvate + ADP' pyruvate + ATP.

For bypass of the Pyruvate Kinase reaction of Glycolysis, cleavage of 2 ~P bonds is required. The free energy change associated with cleavage of one ~P bond of ATP is insufficient to drive synthesis of phosphoenolpyruvate (PEP), since PEP has a higher negative G of phosphate hydrolysis than ATP.

The two enzymes that catalyze the reactions for bypass of the Pyruvate Kinase reaction are the following:

(a) Pyruvate Carboxylase (Gluconeogenesis) catalyzes: pyruvate + HCO3 + ATP' oxaloacetate + ADP + Pi

(b) PEP Carboxykinase (Gluconeogenesis) catalyzes: oxaloacetate + GTP' phosphoenolpyruvate + GDP + CO2

Contributing to spontaneity of the two-step pathway are the following:

- Free energy of cleavage of one ~P bond of ATP is conserved in the carboxylation reaction.Spontaneous decarboxylation contributes to spontaneity of the 2nd reaction (PEP synthesis).

Cleavage of a second ~P bond of GTP also contributes to driving synthesis of PEP.

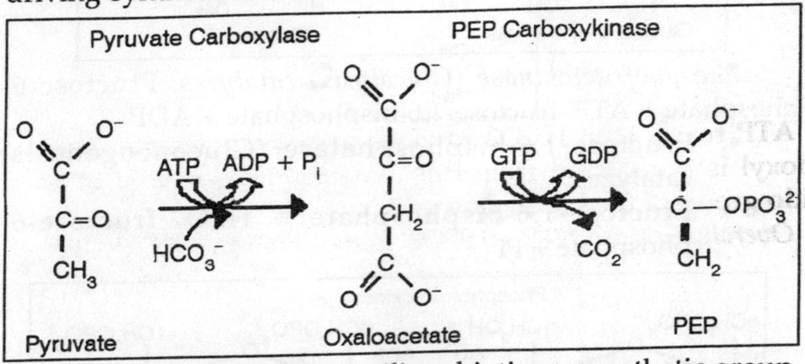

Pyruvate Carboxylase
PEP Carboxykinase
ATP ADP + P_i
HCO_3^-
GTP GDP
CO_2
Pyruvate
Oxaloacetate
PEP

Pyruvate Carboxylase utilizes biotin as prosthetic group. See diagram p. 846 of Biochemistry, 3rd Edition, by Voet and Voet.

Biotin has a 5-carbon side chain whose terminal carboxyl is in an amide linkage to the -amino group of alysine of the enzyme.

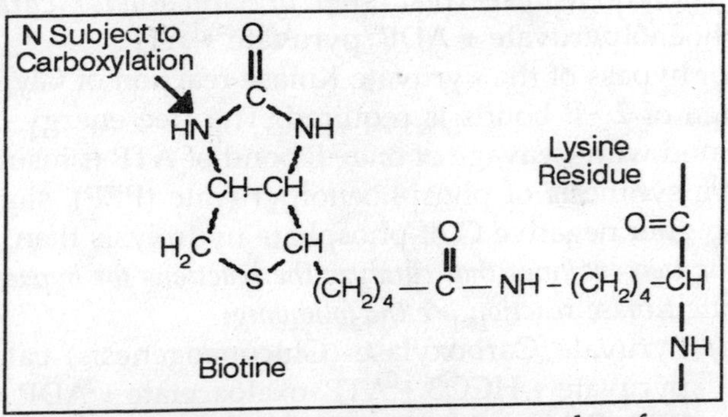

N Subject to Carboxylation
Lysine Residue
Biotine

The biotin and lysine side chains together form a long swinging arm that allows the functional group of biotin to swing back and forth between two active sites.

Biotin carboxylation is catalyzed at one active site of Pyruvate Carboxylase.

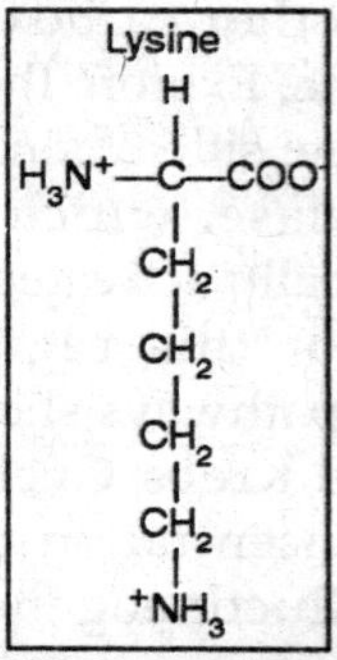

Lysine

$$H_3N^{+}-\overset{\displaystyle H}{\underset{\displaystyle CH_2-CH_2-CH_2-CH_2-{}^{+}NH_3}{C}}-COO^{-}$$

ATP reacts with HCO3 to yield carboxyphosphate. The carboxyl is transferred from this ~P intermediate to N of a ureido group of the biotin ring system.

Overall: biotin + ATP + HCO3' carboxybiotin + ADP + Pi.

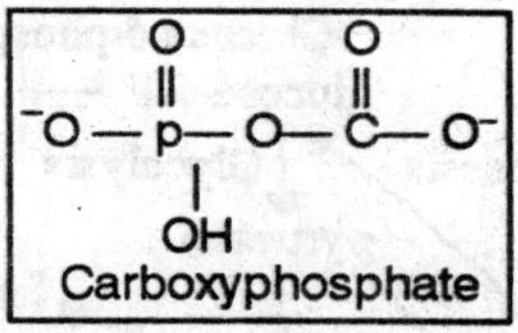

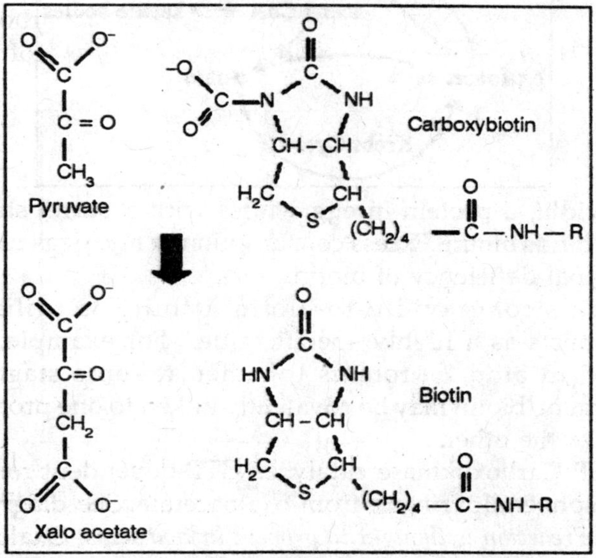

$$^{-}O-\underset{\displaystyle OH}{\overset{\displaystyle O}{\overset{\|}{P}}}-O-\overset{\displaystyle O}{\overset{\|}{C}}-O^{-}$$

Carboxyphosphate

Pyruvate

Carboxybiotin

Xalo acetate

Biotin

biotinyl domain of another carboxylase enzyme, Acetyl-Coenzyme A Carboxylase. Explore the structure of biotin and its attachment via a lysine side chain.

Pyruvate Carboxylase, which converts pyruvate to oxaloacetate, is allosterically activated by acetyl coenzyme A. The adaptive value of this regulation relates to the interconnectness of the pathways shown at right.

Acetyl CoA enters Krebs Cycle by condensing with oxaloacetate, whose concentration tends to be limiting for Krebs Cycle. When Gluconeogenesis is active in liver, oxaloacetate is diverted to form glucose (via PEP). Oxaloacetate depletion hinders acetyl CoA entry into Krebs Cycle. The resulting increase in [acetyl CoA] activates Pyruvate Carboxylase to synthesize more oxaloacetate.

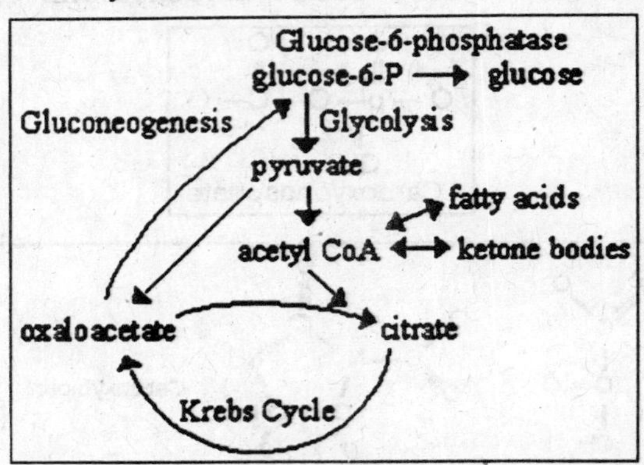

Avidin, a protein in egg whites with a barrel structure, tightly binds biotin. Excess consumption of raw eggs can cause nutritional deficiency of biotin.

The strong avidin-to-biotin affinity is utilized by biochemists as a highly specific"glue." For example, if it is desired to bind 2 proteins together at some stage of an experiment, biotin may be covalently linked to one protein and avidin to the other.

PEP Carboxykinase catalyzes GTP-dependent formation of phosphoenolpyruvate from oxaloacetate. See diagram.

The reaction is thought to proceed in two steps: Oxaloacetate is first decarboxylated to yield a pyruvate enolate anion intermediate. Phosphate transfer from GTP then yields phosphoenolpyruvate (PEP).

PEP Carboxykinase Reaction

oxaloacetate ($^-OOC-CO-CH_2-COO^-$) $\xrightarrow{-CO_2}$ $^-OOC-C(O^-)=CH_2$ $\xrightarrow{GTP \rightarrow GDP}$ $^-OOC-C(OPO_3^{2-})=CH_2$ PEP

In bacterial PEP Carboxykinases, ATP is phosphate donor instead of GTP. In the crystal structure of an E. Coli PEP Carboxykinase at right, pyruvate is at the active site as an analog of phosphoenolpyruvate or oxaloacetate.

A metal ion such as Mn++ is required for the PEP Carboxykinase reaction, in addition to a Mg++ ion that binds with the nucleotide substrate at the active site. The Mn++ is thought to promote the phosphate transfer by interacting simultaneously with the enolate oxygen atom and an oxygen atom of the terminal phosphate of GTP or ATP.

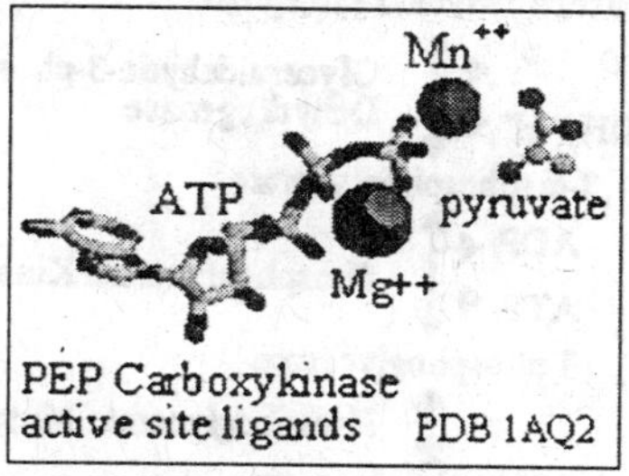

PEP Carboxykinase active site ligands PDB 1AQ2

Explore at right the structure of E. Coli PEP Carboxykinase.

Recommended display options: Select ligand and display as spacefill.

Identify and note relative positions of ATP, pyruvate, Mg++ and Mn++.

We see that the terminal phosphate of ATP would be transferred to the oxaloacetate substrate to yield phosphoenolpyruvate (pyruvate substituting here).

- Select protein, display as cartoon with Colour structure.
- Describe the unique secondary structure of this enzyme.

Now change the display of protein to spacefill, with Colour chain.

GLUCONEOGENESIS INPUTS

The source of pyruvate and oxaloacetate for gluconeogenesis during fasting or carbohydrate starvation is mainly amino acid catabolism. Some amino acids are catabolized to pyruvate, oxaloacetate, or precursors of these. Muscle proteins may break down to supply amino acids. These are transported to liver where they are deaminated and converted to gluconeogenesis inputs.

Glycerol, derived from hydrolysis of triacylglycerols in fat cells, is also a significant input to gluconeogenesis.

Gluconeogenesis pathway is summarized below, with gluconeogenesis enzyme names in red and names of reversible glycolysis enzymes in blue:

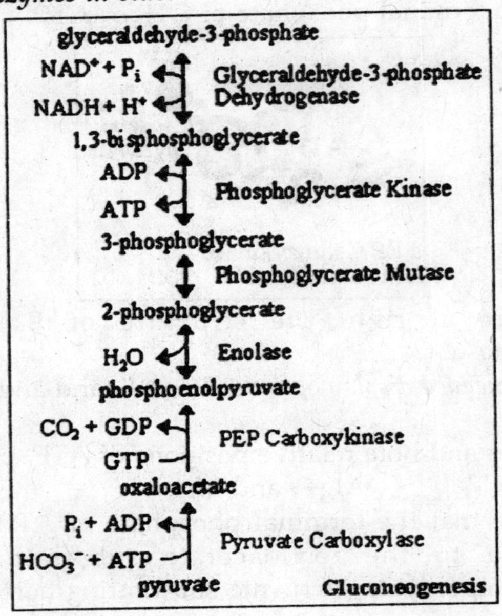

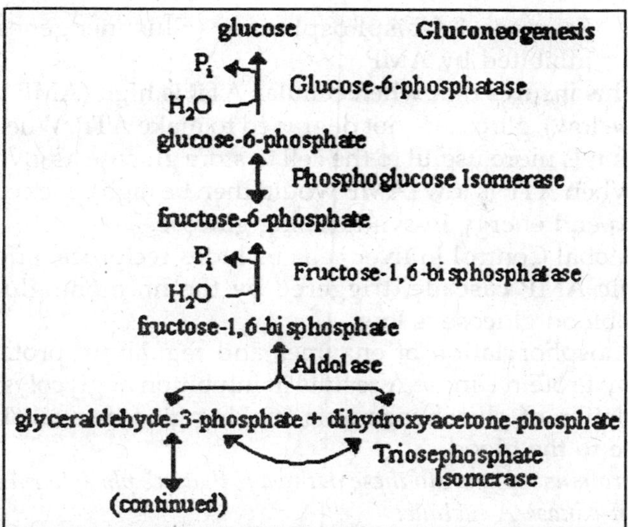

Glycolysis and Gluconeogenesis pathways are both spontaneous. If both pathways were simultaneously active within a cell it would constitute a"futile cycle" that would waste energy. Overall, each pathway may be summarized as follows (ignoring water and protons):

Glycolysis

Glucose + 2 NAD+ + 2 ADP + 2 Pi → 2 pyruvate + 2 NADH + 2 ATP

Gluconeogenesis

2 pyruvate + 2 NADH + 4 ATP + 2 GTP → glucose + 2 NAD+ + 4 ADP + 2 GDP + 6 Pi

Glycolysis yields 2 ~P bonds of ATP.

Gluconeogenesis expends 6 ~P bonds of ATP and GTP.

A futile cycle consisting of both pathways would waste 4 ~P bonds per cycle. To prevent this waste, Glycolysis and Gluconeogenesis pathways are reciprocally regulated. Local Control includes reciprocal allosteric regulation by adenine nucleotides.

- Phosphofructokinase (Glycolysis) is inhibited by ATP and stimulated by AMP.

- Fructose-1,6-bisphosphatase (Gluconeogenesis) is inhibited by AMP.

This insures that when cellular ATP is high (AMP would then be low), glucose is not degraded to make ATP. When ATP is high it is more useful to the cell to store glucose as glycogen.

When ATP is low (AMP would then be high), the cell does not expend energy in synthesizing glucose.

Global Control in liver cells includes reciprocal effects of a cyclic AMP cascade, triggered by the hormone glucagon when blood glucose is low.

Phosphorylation of enzymes and regulatory proteins in liver by Protein Kinase A results in inhibition of glycolysis and stimulation of gluconeogenesis, making glucose available for release to the blood.

Proteins relevant to these pathways that are phosphorylated by Protein Kinase A include:

- Pyruvate Kinase, a glycolysis enzyme that is inhibited when phosphorylated.
- CREB (cAMP response element binding protein) which activates, through other factors, transcription of the gene for PEP Carboxykinase, leading toincreased gluconeogenesis.
- A bi-functional enzyme that makes and degrades an allosteric regulator, fructose-2,6-bisphosphate.

RECIPROCAL REGULATION

- Fructose-2,6-bisphosphate stimulates Glycolysis.
  - Fructose-2,6-bisphosphate allosterically activates the Glycolysis enzyme Phosphofructokinase.
  - Fructose-2,6-bisphosphate also activates transcription of the gene for Glucokinase, the liver variant of Hexokinase that phosphorylates glucose to glucose-6-phosphate, the input to Glycolysis.
- Fructose-2,6-bisphosphate allosterically inhibits the gluconeogenesis enzyme Fructose-1,6-bisphosphatase.

Recall that Phosphofructokinase, the rate-limiting step of the Glycolysis pathway, is allosterically inhibited by ATP. At high concentration, ATP binds to a low affinity regulatory site, promoting the tense conformation. Sigmoidal dependence of reaction rate on [fructose-6-phosphate] is observed at high ATP, as depicted at right.

Phosphofructokinase activity in the presence of the globally controlled allosteric regulator fructose-2,6-bisphosphate is similar to that observed when [ATP] is low. Fructose-2,6-bisphosphate promotes therelaxed state, activating Phosphofructokinase even at relatively high [ATP].

Thus activation by fructose-2,6-bisphosphate, whose concentration fluctuates in response to external hormonal signals, supersedes local control by ATP concentration.

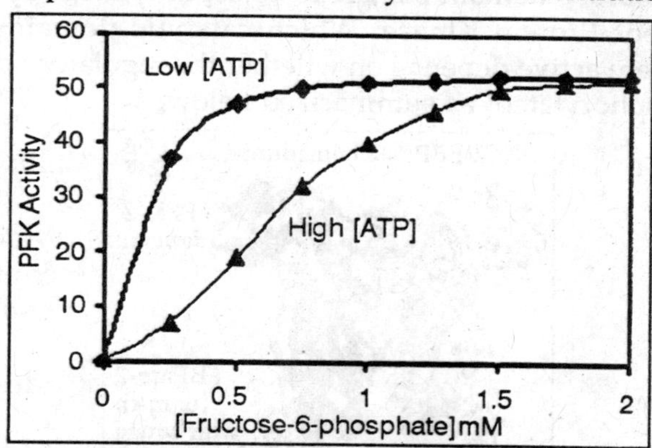

In the Biochemistry Simulations tutorial at right, select the module on Phosphofructokinase, and explore effects of varied concentrations of ATP and the activator fructose-2,6-bisphosphate on the dependence of reaction rate on fructose-6-phosphate concentration.

The allosteric regulator fructose-2,6-bisphosphate is synthesized and degraded by a bi-functional enzyme that includes two catalytic domains:

- Phosphofructokinase-2 (PFK2) domain catalyzes: fructose-6-phosphate + ATP' fructose-2,6-bisphosphate + ADP.
- Fructose-Biosphosphatase-2 (FBPase2) domain catalyzes: Fructose-2,6-bisphosphate + H2O' fructose-6-phosphate + Pi.

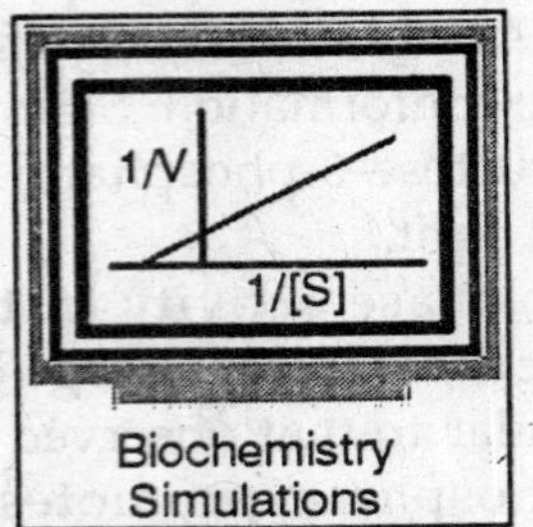

Hold down theControl key while clicking on the above icon. The bi-functional PFK2/FBPase2 assembles into a homodimer. Adjacent to the PFK2 domain in each copy of the liver enzyme is a regulatory domain subject to phosphorylation by cAMP-dependent Protein Kinase. Which catalytic domains of the enzyme are active depends on whether the regulatory domains are phosphorylated, as summarized below.

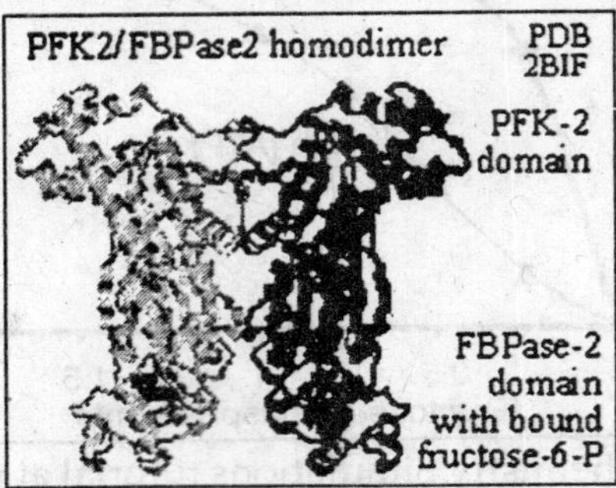

cAMP-dependent phosphorylation of the bi-functional enzyme activates FBPase2 and inhibits PFK2. Fructose-2,6-bisphosphate thus decreases in liver cells in response to a cAMP signal cascade, activated byglucagon when blood glucose is low.

Downstream Effects Include

- Glycolysis slows because fructose-2,6-bisphosphate is not available to activate Phosphofructokinase.

Gluconeogenesis increases because of the decreased concentration of fructose-2,6-bisphosphate, which would otherwise inhibit the gluconeogenesis enzyme Fructose-1,6-bisphosphatase.

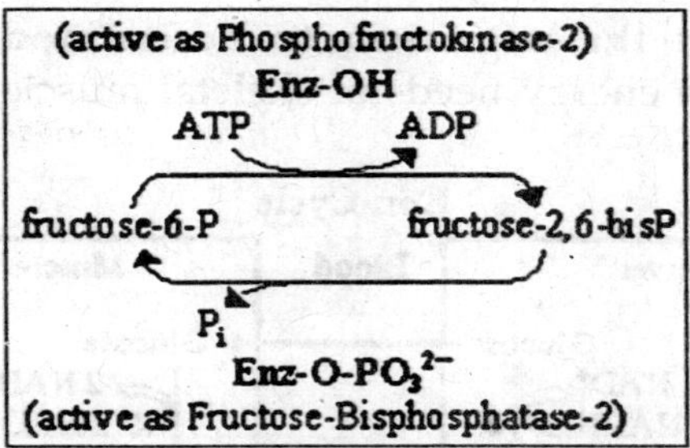

Summarizing effects described above and in the notes on glycogen metabolism, a glucagon-induced cAMP cascade has the following effects in liver tissue:

- Gluconeogenesis is stimulated.
- Glycolysis is inhibited.
- Glycogen breakdown is stimulated.
- Glycogen synthesis is inhibited.

Free glucose formed for release to the blood.

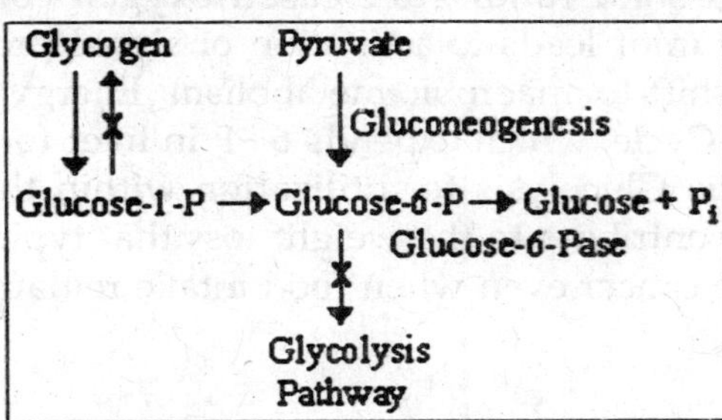

THE CORI CYCLE

For a brief burst of ATP utilization, muscle cells utilize ~P stored as phosphocreatine. Once phosphocreatine is exhausted, ATP is provided mainly by Glycolysis, with the input coming from glycogen breakdown and from glucose uptake from the blood. Lactate produced from pyruvate passes via the blood to the liver where it may be converted to glucose. The glucose may travel back to the muscle to fuel Glycolysis.

The Cori Cycle costs 6 ~P in liver for every 2 ~P made available in muscle.

The net cost is 4 ~P.

Although costly in terms of"high energy" bonds, the Cori

Cycle allows the organism to accommodate to large fluctuations in energy needs of skeletal muscle between rest and exercise.

Cori Cycle

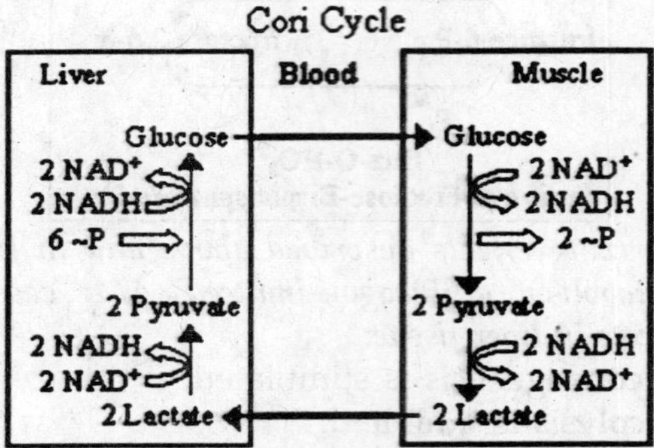

The equivalent of the Cori Cycle also operates during cancer. If blood vessel development does not keep pace with growth of a solid tumor, decreased oxygen concentration within the tumor leads to activation of signal processes that result in a shift to anaerobic metabolism. Energy dissipation by the Cori Cycle, which expends 6 ~P in liver for every 2 ~P produced via Glycolysis for utilization within the tumor, is thought to contribute to the weight loss that typically occurs in iate-stage cancer even when food intake remains normal.

Chapter 10

Postenc Ephalitic Parkinsonism

OVERVIEW

GENERAL OUTLINE

PEP is a progressive neuro-degenerative disease with clinical features referable mainly to the extrapyramidal and oculomotor systems and cognitive deterioration. It represents a chronic complication of encephalitis lethargica (EL) or Economo's encephalitis or sleep sickness that emerged during and after World War I.

SYMPTOMS

Bradykinesia, rigidity, hypomimia, postural instability, gait disorders with falls and sialorrhea.Ophthalmoplegia and oculogyric crises. Elderly patients suffer from continuous deterioration of motor function with dysphagia, incontinence, Levo-Dopa induced psychoses, dystonia, and cognitive impairment.

Neuropathology: multisystem neuronal degeneration and cell loss, with gliosisin many brainstem nuclei, in particular substantia nigra, locus ceruleus, and others with globous neurofibrillary tanglesin residual neurons in many subcortical nuclei.

Cortical pathology is common with frequent NFTs inhippocampus, entorhinal, frontal and insular cortices, differing in distribution from those in AD.

Immunohistochemical studies showed deposits of pathologic tau protein in sucortical fibrillary tangesand astroglia, indicating relationship to tauopathies. Ultrastructure and biochemistry of neuronal and glial deposits are similar to those in Alzheimer diseaase: Ultrastructurally NFTs are composed of 22 nm paired helical filaments and rare 15 nm straiht tubules, Biochemistry of tau protein in PEP shows tau triplet wit h 3 prominent bands at 60, 64, and 68 kDa, differing from tau in PSP, CBD, Pick's disease, and argyrophilic grain disease.

CAREGIVER PROBLEMS

Similar to those in Parkinson's disease and its late complications.

CAUSES AND RISK FACTORS

The causes and risk factors are unknown. Recent microbiological methods were unable to demonstrate any virus, in particular, wild-type influenza virus, that caused influenza pandemic in 1918-21, simultaneously to LE, could not be detected in brain material of PEP patients. Relationship between EL/PEP and influenzy remains enigmatic.

FREQUENCY

Frequency currently extremely rare; since 1970 only single cases reportedMain occurrence of EL between 1918 and 1924. Between 1925-38, PEP reporesented almost 50% of all cases ofParkinsonism diagnosed at that time.

Later on, incidence of PEP ranged from 4 - 30% (mean of 13%) of all cases of parkinsonism. In recent years, incidence of PEP in austopsy series dropped from 6% to almost zero in last decade.

DIAGNOSTIC PROCEDURES

Similar to Parkinson disease (PD), but history of EL necessary. No diagnostic test available.Diff.diagnosis: mainly against PD, but significant differences: 1. onset of symptoms in younger age, including children and adults aged 25-40 yrs; 2. rare occurence of rest tremor; 3. progression of disease in discontinous spurts; 44. prior history of acute EL.

CARE AND TREATMENT

Similar to PD. Patients may variably respond to L-dopa therapy. Late complications to be treated like those in PD.

ONGOING RESEARCH/CLINICAL TRIALS

Trials to identify causing viral agent using modern molecular biologic and genetic methods. Due to almost total extinction of PEP, repository archival material represents the only source for modern investigations, that up to the present, failed to identify a causaive or transmissible agent.

Chapter 11

Oxidative Phosphorylation

DEFINITION

OXIDATIVE PHOSPHORYLATION

The NADH and $FADH_2$ formed in glycolysis, fatty acid oxidation, and the citric acid cycle are energy-rich molecules because each contains a pair of electrons having a high transfer potential.

When these electrons are used to reduce molecular oxygen to water, a large amount of free energy is liberated, which can be used to generate ATP. Oxidative phosphorylation is the process in which ATP is formed as a result of the transfer of electrons from NADH or $FADH_2$ to O_2 by a series of electron carriers.

This process, which takes place in mitochondria, is the major source of ATP in aerobic organisms (Figure below). For example, oxidative phosphorylation generates 26 of the 30 molecules of ATP that are formed when glucose is completely oxidized to CO_2 and H_2O.

Oxidative phosphorylation is conceptually simple and mechanistically complex. Indeed, the unraveling of the mechanism of oxidative phosphorylation has been one of the most challenging problems of biochemistry.

The flow of electrons from NADH or $FADH_2$ to O_2through protein complexes located in the mitochondrial inner membrane leads to the pumping of protons out of the mitochondrial matrix. The resulting uneven distribution of protons generates a pH gradient and a transmembrane electrical potential that creates aproton-motive force. ATP is synthesized when protons flow back to the mitochondrial matrix through an enzyme complex. Thus, the oxidation of fuels and the phosphorylation of ADP are coupled by a proton gradient across the inner mitochondrial membrane (Figure).

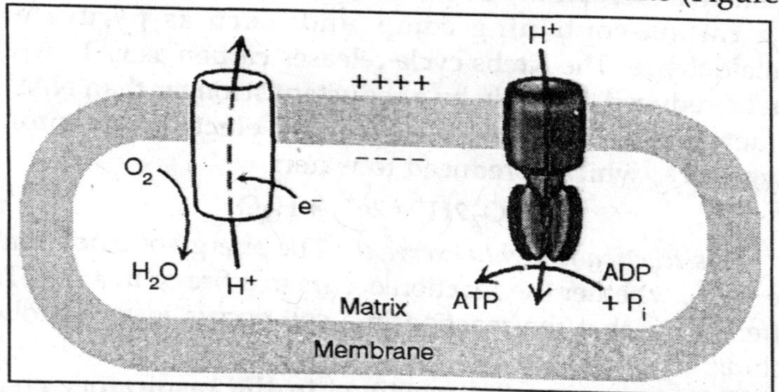

Oxidative phosphorylation is the culmination of a series of energy transformations that are called cellular respiration or simply respiration in their entirety. First, carbon fuels are oxidized in the citric acid cycle to yield electrons with high transfer potential.

Then, this electron-motive force is converted into a proton-motive force and, finally, the proton-motive force is converted into phosphoryl transfer potential. The conversion of electron-motive force into proton-motive force is carried out by three electron-driven proton pumps-NADH-Q oxidoreductase, Q-cytochrome c oxidoreductase, and cytochrome c oxidase.

These large transmembrane complexes contain multiple oxidation-reduction Centres, including quinones, flavins, iron-sulfur clusters, hemes, and copper ions. The final phase of oxidative phosphorylation is carried out by ATP synthase, an ATP-synthesizing assembly that is driven by the flow of protons back into the mitochondrial matrix. Components of this remarkable enzyme rotate as part of its catalytic mechanism.

Oxidative phosphorylation vividly shows that proton gradients are an interconvertible currency of free energy in biological systems.

The reducing equivalents from glycolysis, the Krebs cycle, or other catabolic pathways are carried by coenzymes, particularly NAD, and to some extent FAD. The coenzymes then need to be reoxidized so that the coenzymes can be used again. In anaerobic metabolism, the terminal electron acceptor is a carbon-containing compound, such as pyruvate or acetaldehyde. The Krebs cycle releases carbon as CO_2, which can be reduced, but only by a reductant stronger than NADH. In aerobic metabolism, the terminal electron acceptor is oxygen, O_2, which is reduced to water:

$$\frac{1}{2}O_2 2H^+ + 2e^- \rightarrow H_2O$$

This reaction is highly exergonic: The energy of oxidation is the same, whether the reaction occurs in a fire or in a cell. The difference is that the reaction in a cell occurs in a controlled fashion, in small steps.

The purpose of the reactions in the respiratory chain leading from NADH to oxygen is to conserve the energy of oxidation and convert it to ATP. The reducing equivalents from NADH are transferred through a series of membrane-bound proteins, the cytochromes. As the electrons pass through the cytochromes, hydrogen ions, or protons, are released on one side of the membrane, leading to an electrical potential across the membrane.

The protons flow across the membrane, and the energy associated with this electrical flow is converted to ATP. This overall process by which reducing equivalents are used to make ATP is known as oxidative phosphorylation. The process of proton flow leading to ATP synthesis is known as thechemiosmotic mechanism.

Oxidative phosphorylation occurs on membranes. In bacteria, chemiosmotic ATP synthesis occurs at the cytoplasmic membrane. In plant and animal cells, these reactions occur in the mitochondrion, a double-membraned organelle (Figure).

The ancestor of mitochondria was a bacterial cell incorporated into a nucleated cell, which subsequently lost much (although not all) of its DNA. Most mitochondrial proteins are encoded by nuclear DNA. Some respiratory proteins, along with mitochondrial ribosomal RNA and transfer RNAs, are encoded by mitochondrial DNA.

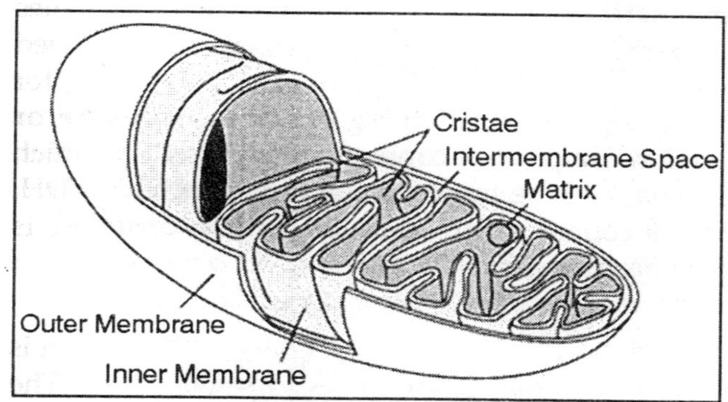

The outer membrane of the mitochondrion contains a large number of pores, so that molecules with molecular weights less than 1,000 can pass from the cytoplasm into the intermembrane space without any specialized transport mechanisms. This means, for example, that NADH, ADP, and inorganic phosphate can reach the intermembrane space from the cytoplasm while NAD and ATP can reach the cytoplasm.

The inner membrane is much less permeable, in part due to the presence of a specialized membrane lipid known as cardiolipin (meaning heart lipid-cardiac cells have a large number of mitochondria). The inner membrane of the mitochondrion is highly folded into cristae, so that it has an interleaved appearance in the electron microscope.

Inside the inner mitochondrial membrane is the space called the matrix. The matrix contains DNA, the apparatus for mitochondrial protein synthesis, and the enzyme systems for the TCA cycle and fatty acid oxidation. (Note that the TCA cycle enzymes are also located in the cytoplasm, because they are involved in many other metabolic transactions in the cell.)

Because generation of NADH for oxidative phosphorylation occurs in the matrix, and because the inner membrane is so impermeable, there must be a large number of specific transport systems to allow small molecules to reach the site where they are catabolized. Finally, many mitochondrial proteins are made by cytoplasmic ribosomes, encoded by nuclear DNA. This means that there must be protein transporting systems to bring these macromolecules into the matrix.

EENERGY OF OXIDATION

The energy of oxidation is given by the redox potential of the reaction. If a piece of copper wire is placed in a solution of zinc sulfate, nothing happens, but if zinc metal is placed in a solution of copper sulfate, the zinc metal is corroded.

Simultaneously, the blue Colour of the copper sulfate solution disappears, and metallic copper is deposited on the surface of the remaining zinc metal:

$$Zn(0) + Cu^{2} \rightarrow Zn^{2-} + Cu(0)$$

The reaction is the transfer of electrons from zinc to copper ions:

$$Zn(0) + Zn^{2+} \rightarrow 2e^{-}$$
$$Cu^{2+} 2^{e-} \rightarrow Cu(0)$$

The overall reaction is the sum of these two half-reactions. The reaction can occur even if the half reactions (that is, the zinc metal and copper ions) are in different containers, as long as the two half reactions are connected electrically, as in Figure.

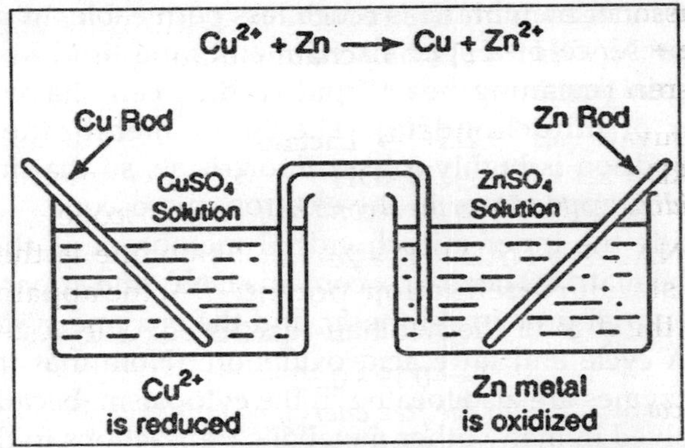

The force associated with electron flow is the voltage. In Figure 2, the voltage can be measured, either by a voltmetre (similar to a battery tester), or by applying an opposing voltage to the wire connecting the zinc and copper electrodes. The amount of voltage required to cancel out the flow of electrons between the electrodes is called the potential of the Zn/Cu cell, designated E. The voltage of a cell depends on the concentrations (more accurately, the activities) of the ions.

Chemists refer reaction energies to a standard state. In this case, the standard state is the one where the ions are present at 1 Molar concentration. The solid metals are always given the activity of 1. In the standard state, the potential of the Zn/Cu cell shown in Fig 11-2 is about 1.1 volts; this means that it can be stopped by application of a direct current from a battery of 1.1 volts opposing the direction of spontaneous electron flow. The reduction potential is related to the standard free energy change associated with reduction by the equation:

$$DG^{0\prime} = n\varpi E^{0\prime}$$

where n is the number of electrons involved in the reaction, ϖ is the Faraday of a constant having the value 23.06 kcal Volts?1 mol?1 (96.5 kJ mol?1 V?1), and Eo? is the standard reduction potential for the two half-reactions. Like other biochemical free-energy changes, the standard reduction potential is determined at pH = 7.0.

The standard reduction potential of a reaction is the sum of the two half-reactions. For example, the standard free energy change associated with the reduction of pyruvate to lactate: Pyruvate + NADH + $H^+ \rightleftarrows$ Lachate + Nad is the sum of the two half-reactions:

$$\text{Pyruvate} + 2e^- + 2H+ \rightarrow \text{Lactate} \qquad \Delta E^{o\prime} - 0.19V$$
$$\text{NAD} + 2e^- + H^+ \rightarrow \text{NADH} \qquad \Delta E^{o\prime} - -0.32V`$$

is oxidized to NAD, so the second half-reaction must be reversed:

$$\text{NADH} \rightarrow \text{NAD} + H^+ + 2e^- \qquad \Delta E^o = +0.32V$$

The standard reduction potential of the reaction is therefore the sum of the two half-reactions:

$$\Delta E^{o\prime} = -0.19 + 0.32 = +0.13$$

This corresponds to the free energy change of:

$$\Delta G^{o\prime} = -(2)(23.06)(+0.13) = -6\,\text{Kcal mol} - 1(-25.1\text{kJmol}^{-1})$$

Biochemical Reduction and Concentration-dependency

This dependence is given by the Nernst equation:

$$E = E^{o\prime} + RT / n\varpi \ln(\Pi[\text{Products}]/\Pi[\text{reactants})$$

In the preceding reaction:

$$E = E^{o\prime} + (2.303RT / n\varpi)\log([\text{lactate}][\text{NAD}]/[\text{Pyruvate}])$$

This is the same form as the dependence of free energy on the concentration of the reductants and products. The term ([H+]/10–7) accounts for the fact that the standard state for biochemical reactions is at pH equal to 7.0.

THE OXIDATION OF NADH

The oxidation of NADH by molecular oxygen provides a large amount of free energy for ATP synthesis. The reduction potential of oxygen is +0.82 V; as can be seen in the preceding formulas, the oxidation of NADH has a standard potential of +0.32 V; therefore, the standard free-energy change associated with the transfer of electrons from NADH to oxygen to make water is:

$$\Delta G^{o\prime} = -n\varpi E^{o\prime} = -(2)(96.5)(0.82 + 0.32 = 220\text{kJ})$$
$$(-53\text{kcal})$$

Compare this with the standard free-energy change of making an ATP from ADP and inorganic phosphate: 31 kJ mol?1 (7.4 kcal mol^{-1}). The purpose of the respiratory chain is to harness this large free-energy change to efficiently synthesize ATP. Three ATP molecules are made by the respiratory chain during the transfer of electrons from NADH to O_2; this corresponds to an efficiency of about 40%. This efficiency is about the same as that of, for example, a diesel engine. Because over 90% of the adenosine nucleotides in the cell are normally fully converted into ATP, the concentration of reactants is lower than standard, and the concentrations of product (ATP) are higher than in the standard state. Thus, cells are able to convert electrical potential into chemical energy at high (although not 100%) efficiency.

A conventional view of mitochondrial structure is represented at right. The respiratory chain is embedded in cristae of the inner membrane.

Spontaneous electron transfer through respiratory chain complexes I, III, and IV is coupled to H+ejection from the matrix to the intermembrane space. Because the outer membrane contains large channels, protons in the intermembrane space may equilibrate with the cytosol. Respiration-linked pumping of protons out of the mitochondrial matrix conserves some of the free energy of spontaneous electron transfers as potential energy of an electrochemical H+ gradient.

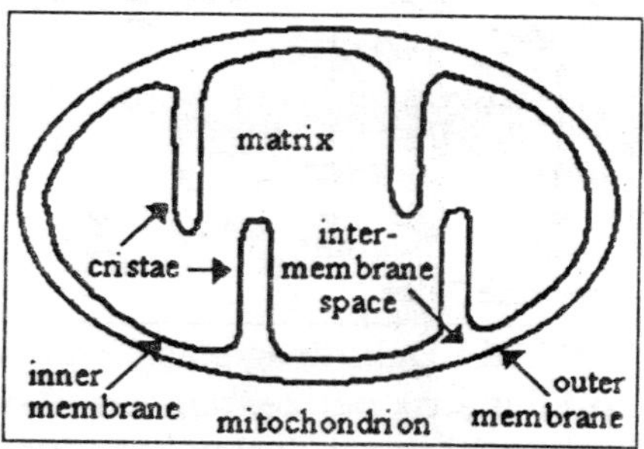

3-D reconstructions based on electron micrographs taken with large depth of field at different tilt angles have indicated that the infoldings of the inner mitochondrial membrane are variable in shape and are connected to the periphery and to each other by narrow tubular regions.

At right is an electron micrograph, provided by Dr. Carmen Mannella of the Wadsworth Centre, of a Neurospora mitochondrion in a frozen sample in the absence of fixatives or stains that might alter the appearance of internal structures.

Tubular cristae connect to the inner membrane via narrow passageways that may limit the rate of H+equilibration between the lumen of cristae and the intermembrane space. There is evidence also that protons pumped out of the matrix spread along the anionic membrane surface and only slowly equilibrate with the surrounding bulk phase, maximizing the effective H+ gradient.

A total of 10 protons are ejected from the mitochondrial matrix per 2 electrons transferred from NADH to oxygen via the respiratory chain. The H+/e ratio for each respiratory chain complex will be discussed separately.

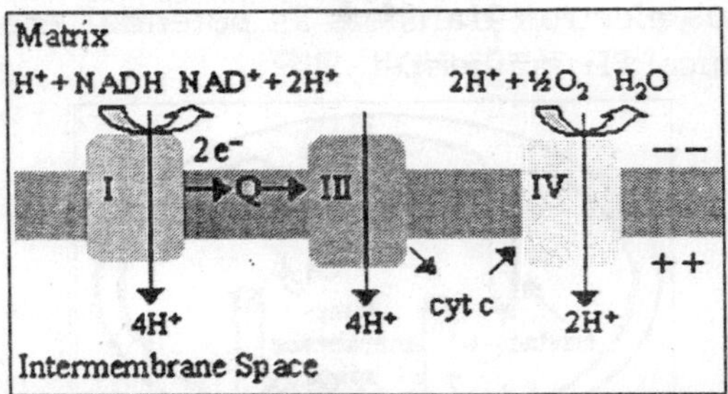

Complex I (NADH Dehydrogenase) transports $4H^+$ out of the mitochondrial matrix per 2e transferred from NADH to coenzyme Q. Lack of high-resolution structural information for the membrane domain of complex I has hindered elucidation of the mechanism of H+ transport through this complex. Direct coupling of transmembrane proton flux and electron transfer is unlikely, because the electron-transferring prosthetic groups, FMN and iron-sulfur Centres, are all located in the peripheral domain of complex I (notes on electron transfer chain). Thus it is assumed that protein conformational changes are involved in H+ transport, as with an ion pump.

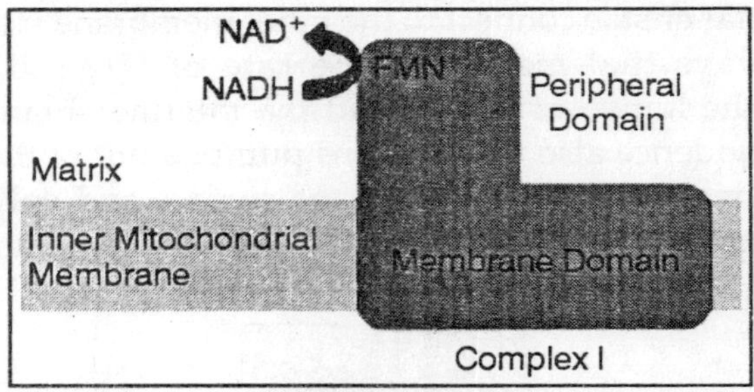

Complex III (bc1 complex): H+ transport in complex III involves coenzyme Q (CoQ). The"Q cycle" depends on the mobility of CoQ within the lipid bilayer. There is evidence for one-electron transfers, with an intermediate semiquinone radical state.

Q Cycle:

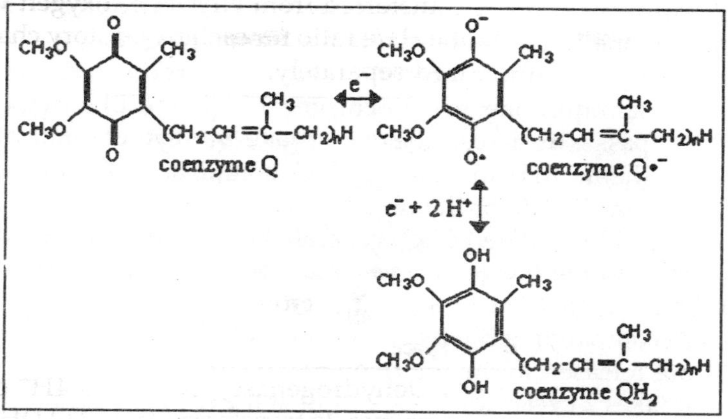

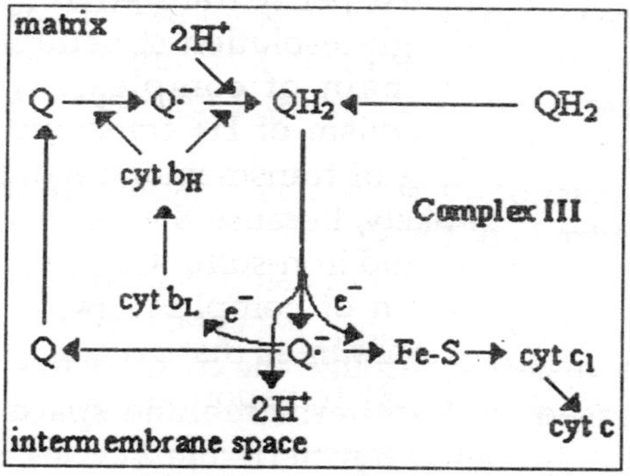

Fig. Q cycle (one version)

In an alternative mechanism that has been proposed, the two electron transfers from QH2 to Fe-S and cyt bL may be essentially simultaneous, eliminating the semiquinone intermediate.

- Electrons enter complex III via coenzyme QH2, which binds at a site on the positive side of the inner mitochondrial membrane, adjacent to the intermembrane space.
- QH_2 gives up one electron to the Rieske iron-sulfur Centre (Fe-S).

Fe-S is reoxidized by transfer of the electron to cytochrome c1, which passes it out of the complex tocytochrome c.

The loss of one electron from QH2 would generate a semiquinone radical, shown here as Q·, although the semiquinone might initially retain a proton as QH·.

A second electron is transferred from the semiquinone to cytochrome bL (heme bL), which passes it across the membrane via cytochrome bH (heme bH) to another CoQ bound at a site on the matrix side of the membrane.

The fully oxidized CoQ, generated as the second electron is passed to the b cytochromes, may then dissociate from its binding site adjacent to the intermembrane space.

Accompanying the two-electron oxidation of bound QH_2, $2H^+$ are released to the intermembrane space.

It takes 2 cycles for CoQ, bound at the site near the matrix side of the membrane, to be reduced to QH^2, as it accepts 2 electrons from the b hemes and 2 H^+ are extracted from the matrix compartment. In 2 cycles, 2 QH^2 enter the pathway, and one is regenerated.

Overall reaction catalyzed by complex III, including net inputs and outputs of the Q cycle:

$$QH_2 + 2H^+(\text{matrix side}) + 2\ \text{cyt c}\ (Fe^{3+})'$$
$$Q + 4H^+_{(\text{outside})} + 2\ \text{cyt c}\ (Fe^{2+})$$

Per $2e^-$ transferred through the complex to cytochrome c, $4H^+$ are released to the intermembrane space. While $4H^+$ appear outside per net 2e transferred in 2 cycles, only $2H^+$ are taken up on the matrix side.

In respiratory chain complex IV, there is a similarly uncompensated uptake of protons from the matrix side ($4H^+$ per O_2 or 2 per $2e^-$).

Thus there are $2H^+$ per 2e that are effectively transported by a combination of complexes III and IV. They are listed with complex III in diagrams (e.g., seeabove) depicting H^+/e^- stoichiometry. Half of the homo-dimeric complex III is depicted at right. The approximate location of the membrane bilayer is indicated. Not shown are the CoQ binding sites near heme bH near heme b_L.

The b hemes are positioned to provide a pathway for electron transfer across the membrane. The protein domain with attached Rieske iron-sulfur Centre (labeled Fe-S) has a flexible link to the rest of the complex. At right, the iron-sulfur Centre protein is Coloured green.

The iron-sulfur Centre changes position during electron transfer. After Fe-S extracts an e from QH_2, it moves closer to heme c_1 (cytochrome c_1) to which it transfers the e.

After the first electron transfer from QH_2 to Fe-S, the CoQ semiquinone is postulated to shift position within the Q-binding site, moving closer to its electron acceptor, heme b_L. This would help to prevent transfer of the second electron from the semiquinone to Fe-S.

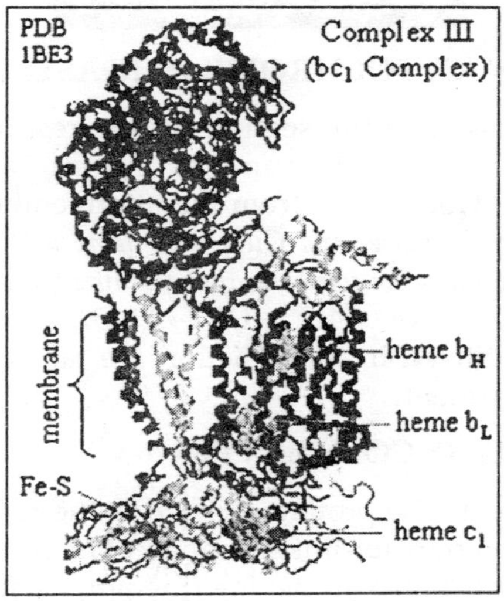

Complex III is an obligate homo-dimer. The iron-sulfur Centre in one half of the dimer may interact with bound CoQ and heme c1 in the other half of the dimer.

At right, arrows indicate the positions of:

- Fe-S in the half of the complex whose proteins are Coloured white/grey.
- Heme c1 in the half of the complex whose proteins are Coloured in shades of blue or green.

Crystal structures on which these diagrams are based, were solved by S. Iwata *et al*, in 1998.

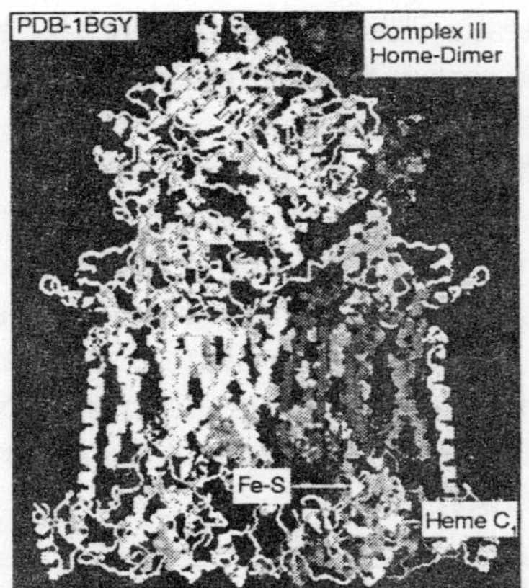

COMPLEX IV (CYTOCHROME OXIDASE)

As discussed in the section on the respiratory chain, electrons are donated to complex IV, one at a time, by cytochrome c, which binds from the intermembrane space.

Each electron passes via CuA and heme a to the binuclear Centre, buried within the complex, that catalyzes oxygen reduction:

$$4e^- + 4H^+ + O2 \rightarrow H_2O.$$

Protons utilized in this reaction are taken up from the matrix compartment.

H+ PUMPING BY COMPLEX IV

In addition to the protons utilized in the reduction of O_2, there is electron transfer-linked transport of 2H+ per $2e^-$ ($4H^+$ per 4e) from the matrix to the intermembrane space.

Structural and mutational studies indicate that protons pass through complex IV via chains of groups subject to protonation/deprotonation, called"proton wires." These consist mainly of chains of buried water molecules, along with amino acid side-chains, and propionate side-chains of the hemes.

Separate H^+-conducting pathways link each side of the membrane to the buried binuclear Centre where O_2 reduction takes place. These include two proton pathways, designated"D" and"K" (named after constituent Asp and Lys residues) extending from the mitochondrial matrix to near the binuclear Centre deep within complex IV. a webpage maintained by the Institute of Biological Information Processing in Germany, a webpage maintained by A. Crofts.

A switch mechanism controlled by the reaction cycle is proposed to effect transfer of a proton from one half-wire (half-channel) to the other. There cannot be an open pathway for H+ completely through the membrane, or oxidative phosphorylation would be uncoupled. (Pumped protons would leak back; see below). The process of switching may involve conformational changes, and oxidation/reduction-linked changes in pKa of groups associated with the catalytic metal Centres. Detailed mechanisms have been proposed

A simplified animation of the entire respiratory chain, at right, includes a simplified representation of the Q cycle. A total of 20 H+is shown being transported out of the matrix, per 4 e transferred from 2 NADH to O_2 (10 H+ per ½ O_2).

Not shown are OH ions that would accumulate in the matrix, as protons, generated by dissociation of water (H_2O↔' H^+ + OH), are pumped out. Also not depicted is the effect on the pH gradient of buffering. The ATP synthase, which is embedded in cristae of the inner mitochondrial membrane, includes the following major subunits:

- F_1: The catalytic subunit, made of 5 polypeptides with stoichiometry.
- F_o: A complex of integral membrane proteins that mediates proton transport.

The F1Fo complex couples ATP synthesis to H+ transport into the mitochondrial matrix. Transport of at least 3H+ per ATP synthesized is required, as estimated from a comparison of the following (calculated in the Tutorial):

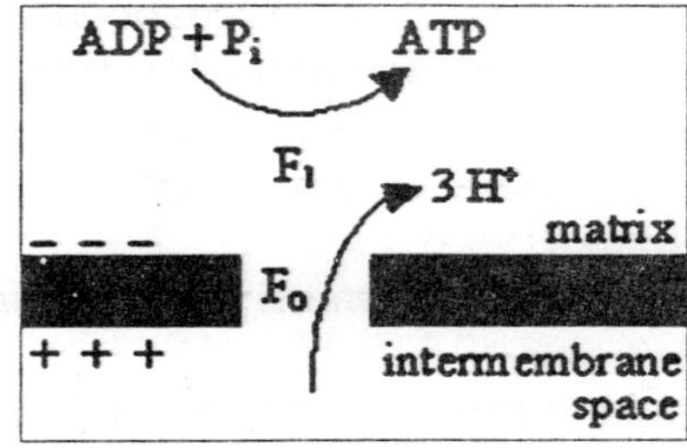

Fig. ATP Synthesis with ΔpH and Δψ

- The free energy change (G) associated with synthesis of ATP under cellular conditions (the free energy required).
- the free energy change (G) associated with transport of each H+ into the mitochondrial matrix, based on the electrochemical H+ gradient (the free energy available per H+).

ATP Synthase structure and mechanism are discussed in more detail elsewhere. The Chemiosmotic Theory of oxidative phosphorylation, for which Peter Mitchell received the Nobel prize is summarized in the diagram at right.

The Chemiosmotic Theory states that coupling of electron transfer to ATP synthesis is indirect, via a H+ electrochemical gradient:

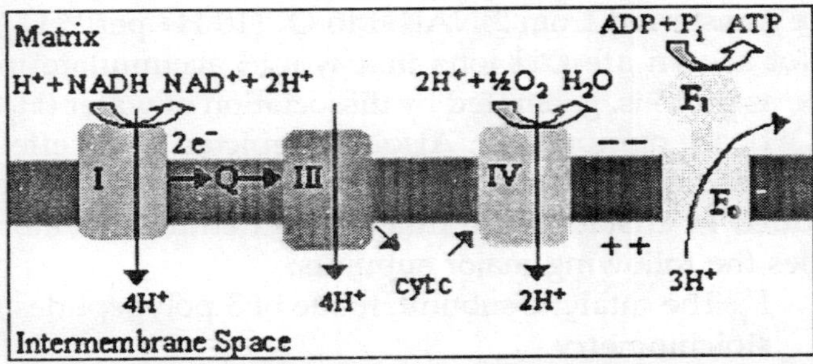

- *Respiration*: Spontaneous electron transfer through complexes I, III, and IV is coupled to non-spontaneous H^+ ejection from the mitochondrial matrix. H^+ejection creates a membrane potential ($\Delta\Psi$, negative in the matrix) and a pH gradient (ΔpH, alkaline in the matrix).
- *F_1F_o ATP Synthase*: Non-spontaneous ATP synthesis is coupled to spontaneous H+ transport into the matrix compartment. The pH and electrical gradients created by respiration are together the driving force for H+ uptake.

Return of protons to the matrix via F_o"uses up" the pH and electrical gradients.

ATP produced in the mitochondria must exit to the cytosol to be used by transport pumps, kinases, etc.ADP and Pi, arising from ATP hydrolysis in the cytosol, must re-enter the mitochondria to be converted again to ATP.

Two carrier proteins in the inner mitochondrial membrane are required for this metabolic cycle. The outer membrane is considered to be not a permeability barrier. The large VDAC channels in the outer membrane are assumed to allow passage of adenine nucleotides and P_i.

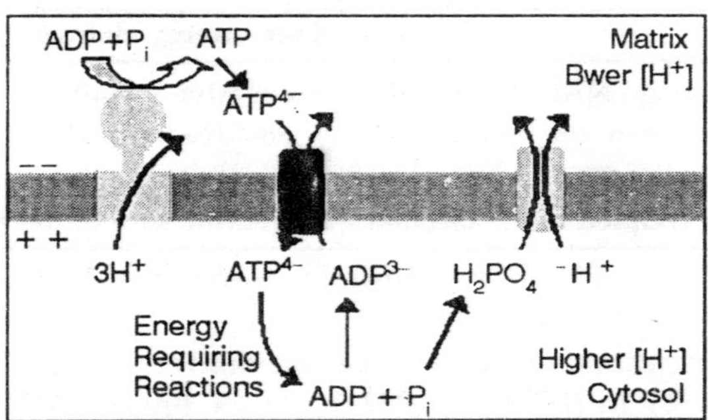

- The Adenine Nucleotide Translocase (ADP/ATP carrier) is an antiporter that catalyzes exchange of ADP for ATP across the inner mitochondrial membrane (p. 496). At cellular pH, ATP has four negative charges, while ADP has 3 negative charges. ADP^{3-}/ATP^{4-} exchange is driven by, and uses up, the membrane potential generated by respiration (one charge per ATP).
- Phosphate reenters the mitochondrial matrix with H^+, by an electroneutral symport mechanism. P_i entry is driven by and uses up the pH gradient (equivalent to one mole of H^+ per mole of ATP).

Thus the equivalent of one mol of H^+ enters the matrix with ADP/ATP exchange and P_i uptake. Assuming transport of 3 mol H^+ by F_1F_o, a total of $4H^+$ would enter the mitochondrial matrix per ATP synthesized.

At right is an animation depicting functioning of the adenine nucleotide translocase and phosphate symport, along with the ATP synthase.

The phenomenon of respiratory control is the subject of today's studio exercise. An oxygen electrode may be used to record [O_2] in a closed vessel. Electron transfer, e.g., from NADH to O_2, is monitored by recording the rate of disappearance of O_2.

Chemiosmotic explanation of respiratory control: Electron transfer is obligatorily coupled to H+ ejection from the matrix. Whether this coupled reaction is spontaneous depends on the pH and electrical gradients.

Reaction	**Free energy change**
e transfer (e.g., NADH to O2)	a negative value*
H+ ejection from the matrix	a positive value that varies with
the H+ gradient**	
e transfer coupled to H+ ejection	Algebraic sum of the above

* $G^{o\prime} = nF\Delta E^{o\prime} = 218$ kJ/mol, for transfer of 2 e from NADH to O_2.

** *For ejection of one H+ from the matrix:*

$\Delta G = RT \ln ([H^+]_{cytosol}/[H^+]_{matrix}) + F\Delta\psi = 2.3\ RT\ (p_{Hmatrix}\ p_{Hcytosol}) + F\Delta\psi$

In the absence of ADP, H+ cannot flow back to the matrix through F_o. The pH and electrical gradients (pHand $\Delta\psi$) are maximal.

As respiration with outward H^+ pumping proceeds, the free energy change for H^+ ejection (positive ΔG) increases and approaches the magnitude of that for electron transfer (negative ΔG). When the coupled reaction becomes non-spontaneous, respiration stops. This is referred to as a static head. In fact there is usually a low rate of respiration in the absence of ADP, attributed to H^+ leaks.

When ADP is added, H^+ enters the matrix via F_o, as ATP is synthesized. This reduces the pH and electrical gradients. ΔG of H^+ ejection decreases. The coupled reaction of electron transfer with H+ ejection becomes spontaneous. Respiration resumes or is stimulated.

Uncoupling reagents (uncouplers) are lipid-soluble weak acids. For example, H^+ (shown in red) can dissociate from the hydroxyl group of the uncoupler dinitrophenol. Diagram p. 834. Uncouplers dissolve in the membrane, and function as carriers for H^+.

Uncouplers block oxidative phosphorylation by dissipating the H+ electrochemical gradient.

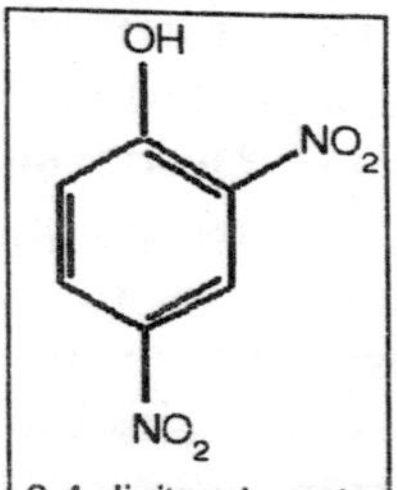

OH

NO_2

NO_2

2,4-dinitrophenol

Protons pumped out are carried by the uncoupler back into the mitochondrial matrix, preventing development of a pH or electrical gradient

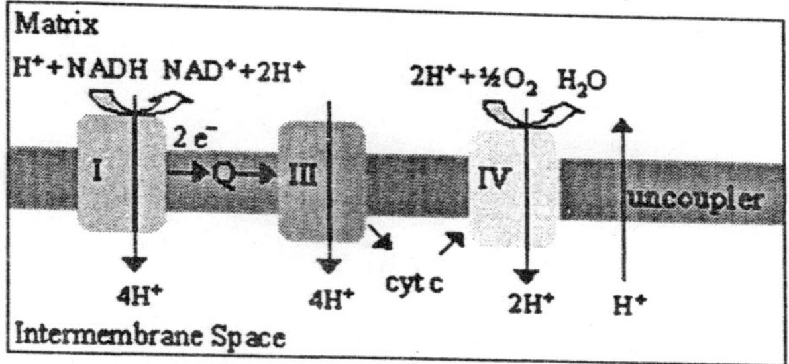

With an uncoupler present there is no pH or electrical gradient. ΔG for H^+ ejection is zero, and ΔG for e^- transfer coupled to H^+ ejection is maximal (spontaneous). Respiration proceeds in the presence of an uncoupler, whether or not ADP is present. Since ΔG for H^+ flux is zero in the absence of a H^+ gradient, and hydrolysis of ATP is spontaneous, the ATP Synthasereaction runs backward in the presence of an uncoupler.

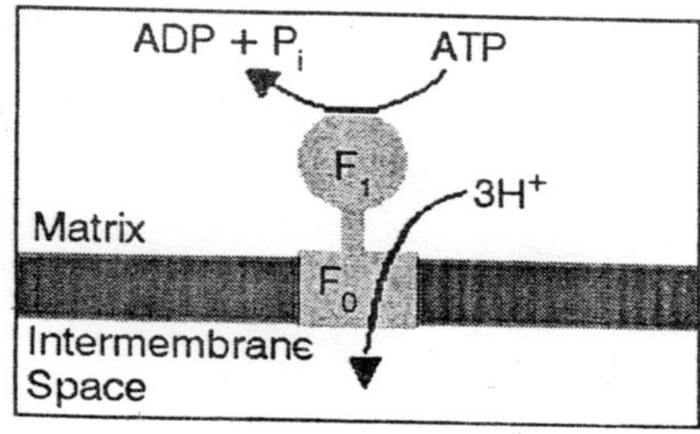

An uncoupling protein (also called thermogenin) is produced in brown adipose tissue of newborn mammals and hibernating mammals. This protein of the inner mitochondrial membrane functions as a H^+ carrier. The uncoupling protein blocks development of a H^+ electrochemical gradient, thereby stimulating respiration. The free energy change associated with respiration is dissipated as heat. This"non-shivering thermogenesis" is costly in terms of respiratory energy unavailable for ATP synthesis, but it provides valuable warming of the organism.

Chapter 12

Disorder/ Diseases of Carbohydrate Metabolism

DISORDERS OF CARBOHYDRATE

Disorders of carbohydrate metabolism occur in many forms. The most common disorders are acquired. Acquired or secondary derangements in carbohydrate metabolism, such as diabetic ketoacidosis, hyperosmolar coma, and hypoglycemia, all affect the central nervous system. Many forms and variants of peripheral nerve disease also are seen in diabetes. The remaining disorders of carbohydrate metabolism are the rare inborn errors of metabolism (ie, genetic defects).

The inherited defects affecting carbohydrate metabolism that have been discovered so far are inherited as autosomal recessive traits. Although multiple affected siblings may be identified in a kindred, most affected individuals are the first identified in a family. Most of the known defects of carbohydrate metabolism appear to be due to point mutations. Whether this will still hold true after the discoveries that the Human Genome Project will make possible is not known.

Most of the inherited disorders of carbohydrate metabolism fall into a few broad clinical syndromes, which are classified by age of onset as follows:

- Infants and early childhood.
  - Episodic lactic acidosis from early infancy, failure to thrive, and hypotonia with or without features that may suggest specific defects.
  - Infantile or early childhood mental retardation, hypotonia, failure to thrive, and other features.

  - Mental retardation/developmental delay with features that suggest storage disorders.
  - Episodic vomiting in infants or young children.
- Childhood or adolescence.
  - Episodic acidosis in childhood or early adulthood, often with features of a specific disorder.
  - Intermittent or episodic ataxia in childhood and adolescence
- In adults and, more rarely, in adolescents or older children.
  - Cramps and weakness, often with some episodes of muscle breakdown and myoglobinuria, during or after heavy anaerobic exercise.
  - Symmetrical neuromuscular disease with weakness and wasting of proximal muscles.

PATHOPHYSIOLOGICAL DISORDERS

The brain, heart, skeletal muscle, and liver depend on ketone bodies or on glucose for energy and for carbon chains to synthesize cellular proteins.

Structure	Metabolite	Enzyme
$CH_3-CH(OH)-CH_2-C(=O)-OH$	Beta-hydroxy-butyrate (3 hyroxy- butyrate)	
		3Hyroxy- Butyrate Dehydrogenase
$CH_3-C(=O)-CH_2-C(=O)-OH$	Acetoacetate	
		Acetcaetate-CoA Synthetase
$CH_3-C(=O)-CH_2-C(=O)\sim CoA$	Acetoacetyl ~ CoA	
		Thiolase
$CH_3C(=O)\sim CoA$ $CH_3C(=O)\sim CoA$	2Molecules of Actyl ~/CoA	

Fig. Ketone Metabolism

For infants as well as individuals with decreased sugar intake because of starvation or dietary choices, ketones are the preferred fuel and source of carbon chains. Nervous and muscle tissues preferentially oxidize ketones over fatty acids, glucose, or amino acids under normal conditions.

KETONE METABOLISM

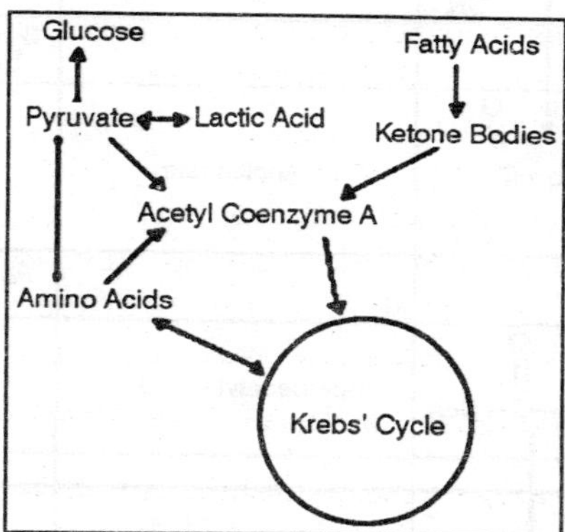

Krebs Cycle

Ketone bodies are largely beta-hydroxybutyrate and acetoacetate. These are formed in the liver and kidneys from the breakdown of fatty acids and a few amino acids. The brains of adults who live on a high-carbohydrate diet lose the ability to synthesize the key enzymes needed to metabolize ketones. These include beta-hydroxybutyrate dehydrogenase and acetoacetyl coenzyme A (CoA) synthetase. For the brain to resynthesize these enzymes to effective levels, starvation for 3-7 days or a high-fat diet is needed.

Neural tissue has far more metabolic activity than other tissues of the body. The brain comprises only a small Percentage of total body weight-approximately 1.4 kg in an average 70-kg person-but uses 20% of the body's oxygen at rest. In addition, neurons (and glia) are constantly synthesizing a variety of neurotransmitters, proteins for axonal flow, and proteins and lipids for regeneration of synaptic vesicles and other components of membranes. Many of these chemicals are synthesized in the brain, in part or wholly from glucose or ketone bodies.

Structure	Metabolite	Enzyme
$CH_3—CH(OH)—CH_2—C(=O)—OH$	Beta-hydroxy-butyrate (3 hyroxy- butyrate)	
		3-Hyroxy- Butyrate Dehydrogenase
$CH_3—C(=O)—OH_2—C(=O)—OH$	Acetoacete	
		Acetoacetate - CoA Synthetase
$CH_3—C(=O)—CH_2—C(=O) \sim CoA$	Acetoacetyl~ CoA	
		Thiolase
$CH_3C(=O) \sim CoA$ $CH_3C(=O) \sim CoA$	2 Molecules of Acetyl ~ CoA	

Fig. Ketone Metabolism

Nonglucose carbohydrates (eg, galactose, mannose, inositol) are clinically important in fetal and neonatal nutrition; however, little is known about their metabolism in the neonate. Glucose and galactose increase postprandially and several other carbohydrates contained in milk do not. Brown *et al*, in their attempt to determine whether postprandial changes in plasma carbohydrate and sugar alcohol concentrations are affected by clinical variables such as postnatal age, milk type, feeding volume, or feeding duration in term newborns, found that galactose is almost cleared completely by the neonatal liver. This provides novel evidence of a marked difference in plasma galactose concentrations [Gal] between relatively younger and older neonates.1

Acute hypoglycemia, as in an alcoholic or a diabetic who has taken too much insulin, causes the brain to be deprived rapidly and abruptly of blood glucose after it has depended for a long time on glucose rather than ketones as its fuel. Beta-hydroxybutyrate dehydrogenase and acetoacetyl CoA synthetase activity in the brain is insufficient to enable the blood ketones to sustain the brain's metabolism. Without fuel to burn, the neural tissue cannot carry out the high level of metabolism needed to pump ions, fire messages, or transmit intercellular chemical messages, let alone repair itself. The groggy state of sudden memory loss (ie, disorientation) and other intellectual deficits that internists tend to call confusion sets in rapidly with acute hypoglycemia, followed just as rapidly by a downhill course into obtundation, coma, and death. If the average adult brain can withstand only 4-6 minutes of oxygen deprivation, it can withstand only about 40-45 minutes without fuel.

Most physicians, as medical students or interns, have had experience treating diabetic ketoacidosis. Although the experience tends to cast ketone bodies as toxins, they are not. The acidosis produced by ketone bodies is toxic and is the direct cause of the Kussmaul breathing typical of patients with diabetic ketoacidosis. Patients with diabetes in this state usually have had normal to elevated blood glucose for the previous days or weeks. Enough glucose enters the brain cells during this time to inhibit synthesis of beta-hydroxybutyrate dehydrogenase and acetoacetyl CoA synthetase.

Again, the ketone-utilizing enzymes in the brain are not working at effective levels. By the time frank ketoacidosis sets in, too little insulin is left for the blood glucose to be able to sustain the high metabolic activity of the brain. The ketone bodies accumulate but cannot provide energy or carbon for the brain. The hydrogen ion associated with the ketones also accumulates, thereby stimulating (or overstimulating) acid receptors in the floor of the fourth ventricle and interfering with reuptake of potassium in the renal tubules.

The coma of hyperosmolar diabetic disorder seems to be related directly to the addition of large concentrations of glucose in the serum, the cerebrospinal fluid (CSF), the extracellular fluid of the brain and, perhaps to some degree, the intracellular fluid as well. The cause of high concentrations of glucose in these patients but not of ketone bodies remains unknown. The lack of ketones probably explains the lack of acidosis. CSF concentrations of glucose mirror those of the serum with a delay of approximately 2 hours. Patients do not regain consciousness immediately when the blood glucose is lowered.

PATHOPHYSIOLOGY OF INHERITED DISORDERS

Many of the clinical features of the inherited disorders of carbohydrate metabolism are caused by the following:

- Lack of glucose for the metabolism of brain, muscle, liver, or kidney (in circumstances in which ketone bodies cannot be used)
- Inability to break down glucose to pyruvate
- Inability to oxidize pyruvate fully in the Krebs cycle

Some features of the inherited diseases are due to excessive storage of abnormal substrates or to normal substrates that cells cannot degrade normally because they lack a specific enzymatic activity. In some diseases, abnormal metabolites (intermediates of pathways that use carbohydrates) block the normal function of the pathway or of related pathways. Defects of the enzymes of the pentose shunt interfere with the normal production of nucleic acids, which are needed by cells as second messengers and as coenzymes of intermediary metabolism, as well as components of RNA and DNA.

Galactose accumulates in the lens of the eye in galactose kinase deficiency.

Metabolism of galactose. The galactose ring differs from glucose in that the hydroxyl group (OH) at carbon-3 is up and not down. To convert galactose to glucose, the ring is first bound to uridine diphosphate (UDP). In this form, the sugar moiety can be epimerized, putting the OH down and forming glucose. The UDP-glucose then is converted to glucose-1-

phosphate, then to glucose-6-phosphate, and subsequently metabolized further.

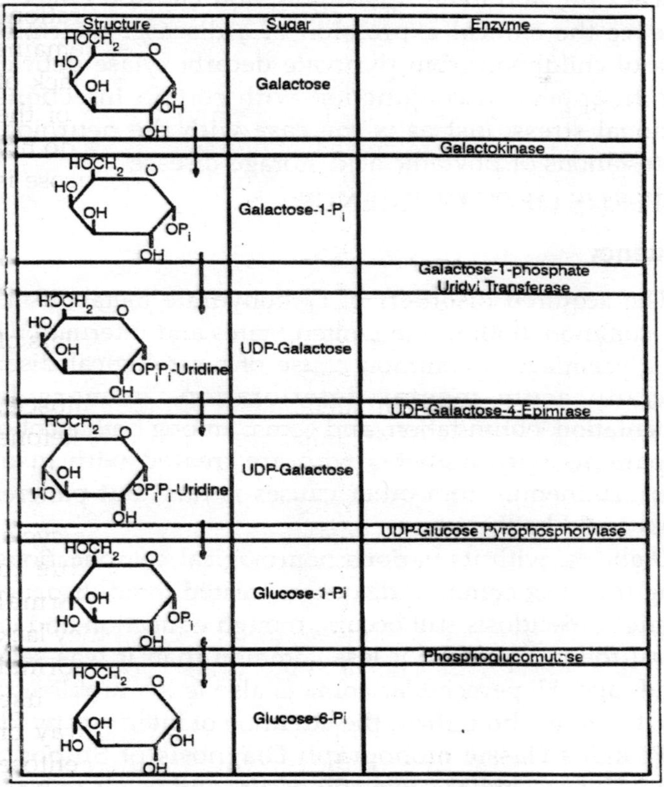

Structure	Sugar	Enzyme
	Galactose	
		Galactokinase
	Galactose-1-P_i	
		Galactose-1-phosphate Uridyl Transferase
	UDP-Galactose	
		UDP-Galactose-4-Epimrase
	UDP-Galactose	
		UDP-Glucose Pyrophosphorylase
	Glucose-1-Pi	
		Phosphoglucomutase
	Glucose-6-Pi	

Fig. Metabolism of Galactose

The osmotic pressure generated causes water to enter the lens; this disruption leads to cataract formation. A similar mechanism may play a role in the pathophysiology of diabetic peripheral neuropathy, which was at one time thought to be due, in part, to the accumulation of sorbitol, a relatively inert but osmotically active metabolite of glucose, and hence to excess water in the cells of peripheral nerves.

Media files 1-8 for the major biochemical pathways involved in carbohydrate metabolism. The expression of a disease often depends on other factors in addition to the causative agent. In inherited disease, variations in the other 30,000 genes, varying diets, infectious disease, access to medical care, and even upbringing and cultural effects can influence the clinical expression of a disease. Intermittent ataxia of childhood from pyruvate decarboxylase deficiency tends to appear in conjunction with certain infections or emotional stress, just as is the case with the neurological manifestations of phytanic acid storage disease

PATTERNS OF OCCURRENCE

Frequency

The acquired disorders of carbohydrate metabolism are fairly common, both in the United States and internationally. Hypoglycemia is a common cause of neurological disease, especially acute mental deterioration, memory loss, disorientation, obtundation, and coma, among both alcoholics and patients with diabetes who are treated with insulin. Hyperinsulinemia from other causes is rare, but pancreatic tumors could be the cause.

Diabetes, with its various neurological complications, is among the most common disorders treated in adult patients. Diabetic ketoacidosis still occurs, though education and close medical follow-up make it less common than it was several decades ago. Hyperosmolar coma is also less a problem than when it was first brought to the attention of internists by Plum and Posner's classic monograph Diagnosis of Stupor and Coma.2 Hyperosmolar coma still occurs and needs to be kept in mind while evaluating an obtunded patient.

The inherited disorders of carbohydrate metabolism are rare. Severe defects of the pyruvate dehydrogenase (PDH) complex and the benign chemical anomaly called pentosuria have been reported in very few (2-6) patients.

Patients with inherited disorders of carbohydrate metabolism tend to be seen at tertiary Centres, where specialist teams are available.

The team may comprise pediatricians, pediatric neurologists, biochemical geneticists, psychiatrists, neurologists with a special interest in mental retardation, and neurologists interested in neuromuscular diseases and the more unusual ataxias.

Neurologists in general practice see few, if any, cases of inherited disorders of carbohydrate metabolism in their careers. This depends partly on whether the practice is in an area in which the carrier state for the inherited disorders is more common and partly on whether the neurologist is on the alert for these disorders. As in every area of medicine, inherited carbohydrate disorders that are looked for tend to be found and those that are not looked for tend to be overlooked. Although each of these disorders was described between 1930 and 1990, certainly they were present in the populations of medically sophisticated countries decades, perhaps centuries, earlier.

Glass *et al* performed a retrospective chart review of new pediatric patients seen during 1998 by specialists of the Division of Clinical and Metabolic Genetics of the Hospital for Sick Children in Toronto, the largest such referral Centre in the country. They found that 81% of specific genetic metabolic diagnoses were made within 1 month of being seen in consultation. Four years after initial consultation, a specific diagnosis was not reached in 5% of patients. The authors concluded that the specific diagnosis of inborn errors of metabolism at a major medical genetic referral Centre tended to be made quickly, or never. Some of the causes of delays in diagnosis include the lack of ready access to existing diagnostic Labouratory testing, technical barriers to the identification of specific metabolic or genetic defects, and incomplete knowledge of genetic defects causing inherited metabolic diseases.

Mortality

Hypoglycemia, diabetic ketoacidosis, and hyperosmolar coma are all potentially fatal but potentially curable conditions. Mortality depends on the time taken by the patient or patient's family to present medical attention, the time taken by the doctor to consider the diagnosis and begin correct treatment, the thoroughness of the treatment, and the time taken to diagnose and thoroughly treat other associated medical problems. The patient with acidosis benefits very little if the ketosis is corrected but an underlying cause is missed, for example, by a delayed spinal tap if meningitis or encephalitis happens to be the illness that has tipped the metabolic balance.

Some of the inherited disorders of carbohydrate metabolism are fatal. For some of these conditions, the more profoundly the mutation interferes with enzyme activity, the more likely the disorder will be fatal. Other inherited disorders usually are not fatal if treated early and thoughtfully. Pentosuria is not even a disease in the strict sense; it is more a metabolic curiosity without significant clinical consequences.

In general, patients with inherited carbohydrate disorders that primarily affect the brain or heart tend to die early; those whose disorders affect the liver and kidney die later on; and those whose disorders affect only muscle have lower mortality rates, unless things go very wrong. Some concern exists among physicians who study heat illness in military recruits that subclinical inherited predispositions to disorders of muscle and of RBCs might be among the factors that predispose to heat illness. As heat illness is the major cause of death of military recruits in boot camp, this is an issue that deserves careful study.

Race

No racial predilection has been reported in most inherited disorders of carbohydrate metabolism. One well-known exception is pentosuria, a very rare genetic occurrence of no clinical significance, which is so far confined to Ashkenazi Jews. The pentose excreted in this condition is xylulose.

Sex

No sex predilection has been reported; the known inherited disorders of carbohydrate metabolism are all autosomal recessive.

Age

A general rule exists in inherited diseases. If the degree

of deficiency of an enzyme varies from family to family, symptoms are likely to appear earliest in families in which the mutation virtually inactivates the enzyme activity and latest in families in which the activity is only just below a certain minimum symptomatic threshold.

Profound defects of activity of the PDH complex are associated with lactic acidosis, hypotonia, profound mental retardation, and often death in the first months or first year of life. Somewhat milder defects and vigourous treatment can be compatible with survival into early to mid childhood.

Moderate defects of the PDH complex are associated with ataxia and other symptoms and signs presenting in childhood and compatible with graduation from college and survival at least into the adult years. Data from studies in the 1970s and 1980s raised the question whether mild defects of this complex might be associated with brain diseases whose onset is in adolescence or early adult life and that are compatible with long survival. These studies have not yet been refuted or disproved.

DIAGNOSIS AND TREATMENT OF ACQUIRED DISORDERS

Diabetic ketoacidosis, hyperosmolar coma, and hypoglycemia present as obtundation or coma. The differential diagnosis is as extensive as it is for obtundation and coma in general. The first clinical task is to make sure the patient survives. Checking the airway, breathing, and circulation (ABCs) and correcting any problems with these is always the first step. The second step is to get baseline blood levels for later analysis before metabolically altering the system during treatment. The third step is to give intramuscular (IM) thiamine followed by 50 mL of 50% intravenous (IV) glucose.

Hypoglycemia can be fatal if not reversed in approximately 45 minutes. A glucose load depletes thiamine stores and precipitates Wernicke syndrome; therefore, thiamine should be given first. Then perform a rapid general examination, looking for clues of serious trouble (eg, abnormal patterns of breathing, unusual body or breath odour, jaundice, pink skin, pallor, petechiae, meningismus, abnormalities of major organs). Next, perform a neurological examination focused on variable conditions associated with coma. Fisher and Plum and Posner have written extensively about the neurological examination of comatose patients.

The neurological examination helps clarify whether the problem is focal or systemic (metabolic). If all signs point to a lesion on one side or in one location, the lesion is focal; imaging studies and IV steroids need to be considered and a neurosurgeon consulted. If the signs are diffuse or do not lateralize, the problem most likely is metabolic. Again, Fisher and Plum and Posner have published schemata that can be followed to go through the myriad possibilities in a logical manner. Subtle clues from the history or the examination can be crucial and therefore must be given serious consideration and addressed immediately.

Bhattacharya *et al*, in their pilot study compared uncooked cornstarch with uncooked physically modified cornstarch (WMHM20) in patients with glycogen storage disease types Ia, Ib, and III. The authors reported longer duration of euglycemia and better short-term metabolic control in most of the patients with WMHM20 when compared with cornstarch.4

Diabetes

The acute severe disorders in patients with diabetes are diabetic ketoacidosis (see Diabetic Ketoacidosis) and diabetic hyperosmolar coma (Hyperosmolar Coma). Hypoglycemia from lack of liver glycogen stores (as in alcoholics) or from excess insulin, whether accidental or intentional, also is an acute emergency. The chronic disorders are more common. These include mononeuritis from diabetes, multiple neuropathy (still called mononeuritis multiplex by some) in patients with diabetes, polyneuropathy in patients with diabetes, and so-called diabetic amyotrophy.

Diabetic Ketoacidosis

Even though the diagnosis may be established clinically,

and even though someone experienced in the care of this problem may sometimes treat the ketoacidosis successfully without supporting Labouratory data, confirming the diagnosis by lab studies is still important. Whether the patient will respond easily or with difficulty or what hazards and problems may lie ahead cannot be known at the outset. The extent of Kussmaul breathing is a guide to the severity of acidosis; sudden relief of this breathing, as the patient is being treated, usually means the pH of the blood is back to normal. This can be checked with serial measurements of arterial blood pH (ie, blood gases) while treating the patient.

Ketones and sugars can be demonstrated in the urine and blood by dipstick, but with each repeated step of treatment, sending samples of blood for an accurate analysis by the Labouratory is advisable. Changes on the ECG may suggest hypokalemia, but, again, getting the Labouratory's ongoing help with serial assays of electrolytes, BUN, and blood pH is advisable. The workup and treatment must include any likely coincidental problem (ie, a new illness other than diabetes) that has tipped the metabolic balance.

Diabetic Hyperosmolar Coma and Hypoglycemia

These are the 2 major diagnoses to consider in any patient presenting with obtundation or coma of metabolic rather than structural etiology. Once the comatose patient's ABCs have been secured, several tubes of blood must be drawn for the host of cellular and biochemical analyses a comatose patient may need. Administer 100 mg thiamine IM and then inject an IV bolus of 50 mL of a 50% solution of glucose through a needle of largest possible diametre (the author uses a 15-gauge needle). This amount of glucose is enough to reverse hypoglycemia but represents a mere"drop in the bucket" for a patient with hyperosmolar coma or even diabetic ketoacidosis.

Rapid clinical improvement after a bolus of glucose confirms the diagnosis of hypoglycemia. Patients with hyperosmolar coma do not respond, but the initial blood glucose, as measured in the Labouratory, is very high (ie, usually well over 400 mg/dL), as is the serum osmolality. The CSF glucose corresponds to the blood glucose but with a 2-hour delay; therefore, the CSF glucose remains high even after the blood glucose and electrolytes are brought slowly back to normal. Remember the dangers of central pontine myelinolysis when correcting electrolyte imbalances due to any cause.

The chronic neurological complications of diabetes are due to damage to peripheral nerves. Mononeuritis and multiple neuropathy are both due to infarction of the tiny artery supplying a particular nerve, the vasa nervorum.

Diabetes is a common cause of these problems, and it often presents with infarction of a single nerve, especially of cranial nerves III and VI and of the long mixed nerves of the limbs. When the distal motor nerves of limbs are involved, the diagnosis may be fairly clear. When the proximal motor nerves are primarily involved, the diagnosis may be difficult. The pattern can be that of weakness of the muscles of one thigh with deep pain (ie, diabetic amyotrophy).

Amyotrophy can be a presentation of diabetes. Sometimes, the opposite limb is affected much later. Amyotrophy tends to be self-limiting and improves without any treatment after a few months. The cause of the waxing and waning of symptoms of amyotrophy is unclear.

In some patients, the pattern is of damage to the lumbar plexus, resulting in diabetic plexitis. Again, the reason for the differences in the extent of involvement is unknown.

The differential diagnosis of mononeuritis includes local pressure along the course of the nerve or its roots from trauma, hematoma, entrapment, abscess, or growths, and other causes of infarction of the arteria nervorum, of which a current leading cause is collagen-vascular disease, especially the peculiar form of periarteritis nodosa that is confined largely to the peripheral nervous system (ie, not clinically affecting other organ systems).

These may be suggested by repeated abnormalities of the erythrocyte sedimentation rate or the C-reactive protein or by consistent elevations of antibodies specific for certain collagen-vascular diseases.

Other causes of multiple neuropathy include Lyme disease, sarcoidosis, acquired and inherited amyloidosis, and, very rarely, first attacks of tumacular neuropathy. Most cases of multiple neuropathies are treatable; therefore, this syndrome is one of the few clinical indications for a nerve biopsy should it be needed to establish the diagnosis.

The sural nerve is almost always the nerve of choice. Consult an experienced surgeon to obtain the specimen. Too small a piece, or one that looked liked nerve but was collagen, does no good, and the patient usually will not consent to a second attempt at biopsy.

Many cases of plexitis, whether brachial or lumbar, unilateral or bilateral, are due to infectious or postinfectious problems. These need to be excluded carefully before diabetes is accepted as the cause of plexitis. The problem is complicated by treatment.

For example, the nondiabetic causes of plexitis often respond to high-dose steroids. Most of the author's patients have responded well to pulse-dose steroids for 5 days as currently used for other immune-mediated neurological disorders.

Steroids elevate blood sugars. This means that diabetes needs to be sought and excluded before a therapeutic trial and that a patient with diabetes who is to be given even pulse-dose steroids needs careful monitoring, often with testing of blood glucose 4 times a day for monitoring excessive elevations of glucose during treatment.

In the era before CT scans of the abdomen and pelvis were available, the worsening of what appeared to be a lumbar plexitis with high-dose steroids pointed to the correct problem, pressure palsy from a pelvic abscess.

Polyneuropathy

Diffuse disorders of peripheral nerves exist, both sensory and motor, but often the sensory disorders are the most profound, clinically affecting the longest Fibres first. This is a common problem in neurology in general, and it is also a common problem in people with diabetes.

The patients have numbness and tingling in the toes and feet and, later, up the legs. By the time the sensory loss reaches the knees, the fingertips begin to be involved.

Often, small-Fibre sensation (eg, superficial touch, pinprick, appreciation of hot and cold) is involved early and most predominantly. These patients are likely to scald themselves by getting into too hot a bathtub. They test the water with their feet, but their feet can no longer warn them of the high temperature.

To some extent, the severity of the neuropathy correlates with the severity of the diabetes and with episodes of lack of metabolic control.

To some extent, the neuropathy may improve if control is restored. However, the correlation is not at all tight. Some patients presenting with neuropathy have mild diabetes; in others, the neuropathy occurs and does not get better readily, regardless of the degree of control of blood glucose levels.

Some patients present with a marked dysesthetic pain, others do not. In many patients, later rather than sooner, a marked autonomic component is present, which is not present in others.

When the autonomic system is involved, the distal skin often becomes mottled, cyanotic or pink, smooth, and loses hair follicles to become shiny. Capillary refill may become poor. Postural hypotension occurs in advanced stages of autonomic involvement.

Treatment of complications of the peripheral nervous system in a patient with diabetes is still poor. Good control of blood glucose levels helps prevent or delay onset of these complications; therefore, young patients today often are put on insulin pumps, and blood glucose is monitored much more closely.

The neuropathic pain of diabetes often responds to modern antiepilepsy drugs, as does neuropathic pain of other etiologies. The authors' patients have responded to gabapentin (Neurontin), usually started at very low doses but often increasing to the large doses used in epilepsy. Pain can be well controlled by maintaining trough levels of 6-12 mg/dL.

DIAGNOSIS AND TREATMENT OF INHERITED DISORDERS

Most of the inherited disorders of carbohydrate metabolism fall into a few broad clinical categories.

These are as follows:

- Episodic lactic acidosis in infancy, failure to thrive, and hypotonia with or without features suggesting specific disorders.
- Infantile mental retardation, hypotonia, failure to thrive, and other features
- Mental retardation/developmental delay with features that suggest storage disorders.
- Episodic vomiting in infants or young children.
- Episodic acidosis in childhood or early adult life, often with features that suggest specific disorders.
- Intermittent or episodic ataxia in childhood.
- Cramps and weakness with anaerobic exercise, often with some episodes of muscle breakdown and myoglobinuria, during or after heavy exercise.
- Symmetrical neuromuscular disease with weakness and wasting of proximal muscles.

Episodic lactic acidosis in infancy, failure to thrive, and hypotonia with or without features suggesting specific disorders.

In this category, an infant presents with signs of acidosis, tachypnea, marked irritability, and tachycardia. As the acidosis worsens, the infant becomes obtunded or comatose. Muscle tone is very poor; the infant is flaccid and has poor head control. If the infant is more than a few hours or days old, signs of failure to thrive may be seen: weight loss or at least poor gain in weight, microcephaly, and poor linear growth. Prognosis is poor even with current vigourous metabolic management. Without such management, the infant usually dies in a short time.

In the absence of other clinical features, this may be the presentation of one of several defects of the PDH multienzyme complex, particularly of its E components.

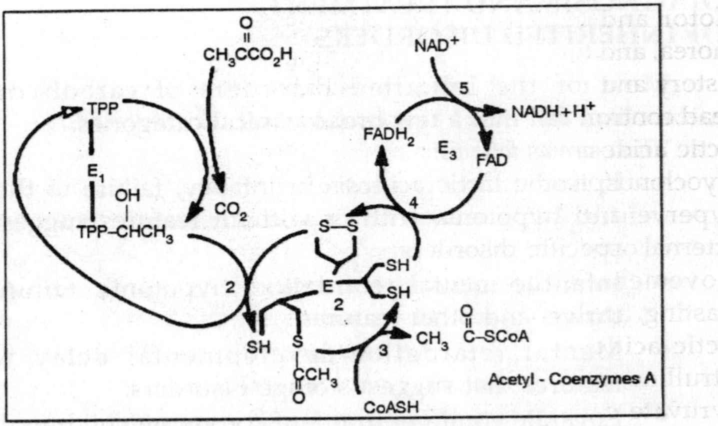

Fig. Pyruvate Dehydrogenase Complex

Scheme of the major reactions of the pyruvate dehydrogenase complex. Initial oxidative decarboxylation of pyruvate is shown as reaction:

- Transfer of the 2-carbon moiety to the lipoyl side chain of lipoyl acetyltransferase (E2) as reaction.
- Formation of acetyl coenzyme A as reaction.
- Oxidation of the lipoyl moiety as reaction.
- Reduction of NAD as reaction.
- The regulatory enzymes acting on pyruvate decarboxylase (E1) are not shown.

The components are called E1 (ie, pyruvate decarboxylase [and the phosphorylase and dephosphorylase that modify the activity of E1]), E2 (ie, lipoyl acetyl transferase), and E3 (ie, lipoamide, also known as dehydrogenase; dihydrolipoamide: NADH oxidoreductase). Several cases of Leigh syndrome appear to have been caused by failure of activation of pyruvate decarboxylase, presumably because of defects in phosphorylation. (Three of the 4 infants described thus far with defects of E3 also had the complex clinical picture described by Leigh.5) In theory, a profound congenital deficiency of thiamine ought to produce episodic lactic acidosis in infants with failure to thrive and hypotonia.

Leigh syndrome (ie, subacute necrotizing encephalomyelopathy) usually begins in the first or second year of life.

Motor and cognitive regression occur along with ataxia, chorea, and tonic spasms. Depending on the age of the patient, history and physical findings include loss of muscle tone and head control, decreased appetite, nausea, vomiting (often with lactic acidosis), irritability to the point of continuous crying, myoclonus, and primarily generalized seizures. Episodes of hyperventilation, apnea and gasps, sobbing without tears, external ophthalmoplegia with pendular or rolling nystagmoid movements of the eyeballs, loss of swallowing, muscle wasting, and loss of reflexes are observed. In cases of postnatal lactic acidosis with elevations in the levels of ammonia, citrulline, and lysine, consider defects of the B-complex of pyruvate carboxylase, the first enzyme in the pathway that generates glucose from fatty acids and amino acids through gluconeogenesis. Death occurs by about 3 months of age.

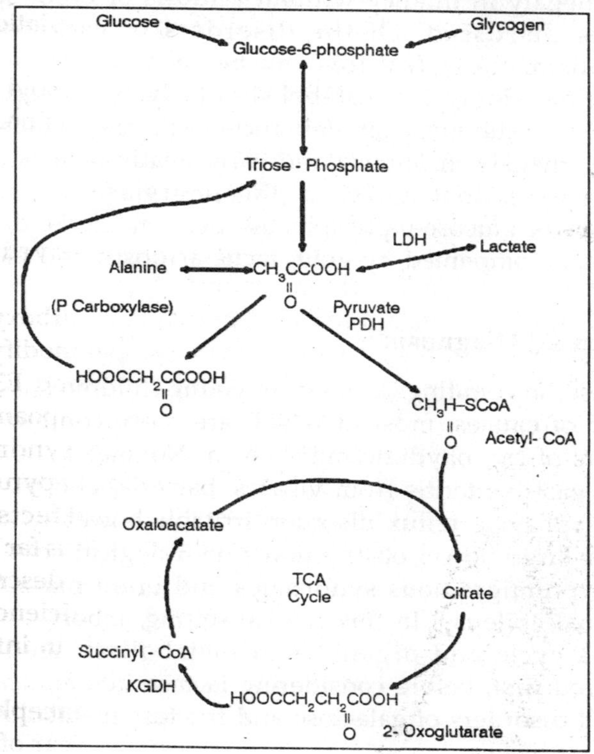

Major metabolic relations of pyruvate. Many of the reactions, such as those between glucose-6-phosphate and pyruvate, pyruvate and alanine, and pyruvate and lactate, are of course reversible. Abbreviations: LDH - lactic dehydrogenase; PDH - pyruvate dehydrogenase complex; p carboxylase - pyruvate carboxylase; TCA cycle - Krebs tricarboxyiic acid cycle; KGDH-a-ketoglutarate dehydrogenase complex.

Mental Retardation/Developmental Delay with Features that Suggest Storage Disorders

The features that suggest storage disorders are hepatomegaly, a large forehead with frontal bossing, and macrocephaly. A large number of diseases, including storage disorders caused by inborn errors of metabolism, can cause hepatomegaly in infancy without acidosis or other specific features discussed. Of the disorders of carbohydrate metabolism, these features can be seen with defects of production, storage, or catabolism of glycogen, especially deficiency of the glycogen debrancher enzyme, which leads to hepatomegaly in infancy and early death; defects of 3 of the 4 proteins that make up phosphorylase kinase; and deficiency of glucose-6-phosphatase (ie, von Gierke disease), which is accompanied by gout, lactic acidosis, and elevated lipids.

Differential Diagnosis

Episodic vomiting in infants or young children has a large number of causes, most of which are more common than disorders of carbohydrate metabolism. Nonmetabolic causes include gastroenteritis from viruses, bacteria, and parasites; gastroesophageal reflux disorder; irritable bowel syndrome; small or large bowel obstructions; neurological conditions, such as premigrainous syndromes and brain masses; and abdominal epilepsy. In this clinical setting, inborn errors of the urea cycle and organic acid metabolism need to be considered first, before considering lactose intolerance from inherited disorders of galactose and fructose metabolism.

Workup and Treatment

Hereditary fructose intolerance and defects of fructose 1, 6-diphosphate aldolase are discussed under Episodic lactic acidosis in infancy, failure to thrive, and hypotonia with or without features suggesting specific disorders earlier in this article. Disorders of galactose metabolism are discussed in the subsection Infants with mental retardation, hypotonia, failure to thrive, and other features.

Episodic Acidosis in Childhood or Early Adult Life

This often presents with features of a specific disorder, including many of the differential diagnoses mentioned under Episodic lactic acidosis in infancy, failure to thrive, and hypotonia with or without features suggesting specific disorders. Specific features may enable rapid diagnosis.

Intermittent or Episodic Ataxia

This category is characterized by the sudden, almost abrupt, onset of ataxia, extrapyramidal symptoms in the upper extremities, and dysarthria soon after or in conjunction with an infectious illness or severe emotional upset. In one patient monitored by one of the authors, death of a grandparent and critical school examinations were sufficient stimuli to initiate attacks of ataxia. Physical findings in this patient also included a high arched palate, minimal kyphoscoliosis, minimal pes cavus, and extra rows of lashes on the upper eyelid. When the boy was 8 or 10 years old, a mild attack was precipitated by an environmental challenge with flashing red lights from scintillation counters and Geiger counters for about 30 minutes in a dimly lit room. For this patient and others, the episodes of ataxia usually last about 2 weeks. Although the neurologic symptoms usually are accompanied by increased serum lactic acid, frank acidosis is not always present.

The underlying defect in this patient was shown to be that of the first enzyme of the PDH complex, E1. Eventually, he was placed on a diet high in thiamine and relatively low in carbohydrates; the number of attacks subsided. While this patient graduated from college, others with similar clinical presentations have had at least a mild degree of mental retardation. Work in the 1970s and early 1980s raised the question whether familial cases of a childhood- or adolescent-onset ataxia (one with slower progression than is seen in classical Friedreich ataxia) was associated with moderate defects of the activity of PDH, E3. The association has not been confirmed. Classic Friedreich ataxia appears to be caused by a single genetic defect and does not appear to be associated with measurable abnormalities in the E3 pathway

Differential Diagnosis

Intermittent or episodic ataxia in childhood and adolescence is a rare condition. Porphyria might induce this, and multiple sclerosis, though rare in young people, certainly can present with ataxia; however, multiple sclerosis almost always manifests in other ways. Postinfectious cerebellar dysfunction usually presents as a single episode that clears in weeks or months and does not recur. Griggs and colleagues have described patients with a familial form of intermittent ataxia, unrelated to pyruvate defects, that responds to treatment with acetazolamide.

Workup

E1 is the first enzyme of the PDH multienzyme complex. Phosphorylation inactivates this pyruvate decarboxylase, which activates by removal of the phosphate. The decarboxylase requires thiamine pyrophosphate as a substrate. Patients deficient in this enzyme have elevations of metabolic by-products related to pyruvate degradation during clinical attacks of ataxia (ie, excess lactic acid and alanine in blood or urine, or both). The primary enzyme defect can be demonstrated in cultured fibroblasts and probably can be demonstrated in WBCs or muscle tissue. The substrate, pyruvate, is so labile and the enzyme activity of this and other components of the PDH complex so sensitive to environmental contaminants that carefully controlled assays need to be performed in a research Labouratory.

Treatment

The young man discussed earlier in detail was treated with a low-carbohydrate diet and high doses of thiamine. Some patients with autosomal recessive ataxia with large-Fibre-posterior column sensory loss, areflexia, cardiomyopathy, and bony stigmata have been treated with some success with a modified (titrated) ketogenic diet. Dietary changes were introduced gradually until moderate to high levels of ketones appeared in the urine without producing frank metabolic acidosis.

Black McArdle first described these symptoms of the disease at Guy's Hospital in London.9 (He was so named to distinguish him from his brother"Red," also a distinguished physician there; the pseudonyms referred to the Colour of their hair.) Type V glycogenosis due to muscle phosphorylase deficiency is the clinically mildest of the glycogen storage diseases, and the symptoms may not occur until adolescence or young adulthood.

The symptoms tend to be induced during exercise when muscle depends on glycolysis rather than oxidative phosphorylation as a source of fuel. Exercise that includes sprinting or vigourous running most likely will induce symptoms.

Often, onset of symptoms comes during what normally would represent the subjective"second wind." Over time, patients may learn to slow down or stop at the first hint of symptoms, and then to begin again, with what has been called a second wind but which is probably a pause in which oxidative metabolism can generate sufficient ATP in muscle so that anaerobic activity can begin again.

If this speculation is true, the"second wind" that is peculiar to myopathies from glycolytic/glycogenolytic defects differs in its nature from the second wind experienced by healthy athletes. The patients therefore can carry on anaerobic exercise in spurts over a long time, presumably with aerobic exercise in between. Affected individuals often can tailor their activities to take a break at the first sign of cramping, and then resume activities shortly thereafter.

Without this pause and resumption in athletes with glycolytic/glycogenolytic defects, muscles suddenly become so weak that the athlete becomes virtually immobile from painful cramps, and the muscles that were exercised are very sore and tender. This disease and the finding of elevated lactic acid levels in the blood after exercise led to the erroneous notion that muscle cramps in general are caused by lactic acidosis. The muscles in the individuals affected with glycolytic defects continue to be very sore even after the cramps resolve.

If the urine is not frankly red, it must be examined to see if it still contains heme groups. In myopathy due to disorders of glycolysis, the heme pigment proves to be myoglobin, not hemoglobin, when detailed analyses are carried out. With McArdle disease, muscle enzymes and creatine levels usually are elevated in the blood after a symptomatic episode. In this condition, muscle histology demonstrates marked increase in glycogen stores in the myofibrils. Histochemical or immunoassays for specific enzymes can augment Labouratory diagnosis by identifying the precise enzyme defect.

The original enzyme defect described by McArdle9 and then by Schmid and Mahler10 was localized to the glycolysis pathway. The deficiency ultimately was identified in muscle phosphorylase.

Through the past decade, MR spectroscopy of forearm muscles *in vivo* has led to the diagnosis. Deficiencies of several other glycolytic enzymes (eg, muscle phospho-fructokinase, phosphoglycerate kinase, phosphoglycerate mutase, lactate dehydrogenase) produce similar clinical pictures and are indistinguishable from muscle phosphorylase deficiency, unless enzymatic tests are performed.

Muscle breakdown occurs in disorders of glycolysis and glycogenolysis during muscle activity that is largely dependent on glycolysis or anaerobic metabolism. This is in striking contrast to the circumstances of muscle breakdown in disorders related to lipid or ketone metabolism. In these disorders, muscle breaks down, with similar symptoms, under circumstances in which muscles are relying on oxidative metabolism, especially of ketone bodies, as a fuel source. Activity most likely to produce symptoms under these circumstances is prolonged, less vigourous exercise over several hours, which is exacerbated by starvation or high-fat, high-protein meals and occurs more often in a cold or cool environment. Histologically, myofibrils may demonstrate increased lipid droplets, but in contrast to the glycogenoses, glycogen stores are either normal or even slightly depleted.

Differential Diagnosis

The differential diagnosis includes all the acute and symptomatic myopathies, including various forms of polymyositis. Many of these appear to be viral or postinfectious; still others appear to be paraneoplastic or associated with motor neuron degenerations or systemic collagen-vascular diseases but not with problems of intermediary metabolism.

The circumstances under which cramps and weakness occur may provide insight into whether a metabolic disorder is present. If evidence of metabolic etiology appears, differentiating between disorders affecting white muscle (ie, the carbohydrate errors discussed here) and disorders affecting red muscle (ie, errors of lipid and mitochondrial metabolism) is essential, because these disorders can present similar pictures but require very different metabolic treatment strategies. Making the correct diagnosis clearly is important since, for each of these groups of disorders of glycolysis/ glycogenolysis, mitochondrial transport, and oxidative phosphorylation, treatment approaches are quite different.

Workup

Prior to the availability of direct enzyme assays, defects of muscle phosphorylase had to be inferred from the clinical features of muscle cramping with vigourous exercise in conjunction with accumulation of lactic acid in the blood and myoglobin in the urine. The forearm ischemia test further substantiated the diagnosis. For this test, a blood pressure cuff is inflated to systolic pressure. With the cuff on, ampules of venous blood are drawn from the antecubital fossa via an indwelling catheter attached to a stopcock before, during, and after 1 minute of forearm exercise (eg, having the patient squeeze a partially inflated blood-pressure cuff repeatedly). In reality, many patients with phosphorylase deficiency cannot continue to exercise for a full minute. Blood samples are assayed for glucose, lactic acid, and sometimes pyruvate. Affected individuals have lower glucose levels and higher lactate and pyruvate levels compared to reference values in blood samples taken 1-2 minutes after exercise.

In the last 25 years, the preferred diagnostic test for McArdle disease has been a phosphorylase assay directly on muscle tissue. Today, histochemical or biochemical techniques can be used to diagnose deficiencies in even the rare enzymes that clinically simulate McArdle disease. Again, these enzymes include muscle phosphofructokinase, phosphoglycerate kinase, phosphoglycerate mutase, and lactate dehydrogenase. In the past 2 decades, diagnosis also has been made by analysis of metabolites *in vivo* on human forearm muscle by means of MR spectroscopy.

Treatment

DiMauro *et al* gave a thorough review of strategies to treat disorders of muscle phosphorylase.11 Pyridoxine, at 50 mg/day, seems to be effective. (Note that at 100 mg/d, this vitamin can produce a painful distal sensory neuropathy.) High-protein diets also were effective, at least for several weeks, but determining the long-term success of dietary treatment will be possible only after adequate clinical trials.

Current therapeutic options include a carbohydrate-rich diet and carbohydrate ingestion shortly before strenuous exercise in combination with medically supervised aerobic training of low to moderate intensity.

Only rarely are cases of progressive neuromuscular diseases caused by defects of glycogen metabolism. When they are, children or young adults generally develop, over months or years, progressive weakness of the limb-girdle muscles and the proximal muscles of the limbs. Patients have no pain under ordinary circumstances, but heavy exercise may induce fatigue and cramps. Symptoms depend on the nature of the underlying metabolic defect. Muscles may or may not be atrophied at presentation but eventually waste. Patients do not complain of muscle tenderness unless cramping occurs. Some patients with debrancher deficiency have evidence of cirrhosis; others become asymptomatic in adult life.

Differential Diagnosis

This condition is essentially indistinguishable from Limb-Girdle Muscular Dystrophy. The symptoms can be demonstrated in a number of acquired diseases, including various forms of polymyositis or inflammatory myopathy, various neuropathic diseases (especially proximal forms of motor neuron disease and the limb-girdle pattern of painless diabetic multiple neuropathy or diabetic plexitis), inherited mitochondrial myopathies (often associated with ragged red Fibres), disorders of lipid metabolism, and/or defects of lipid as carnitine transport at the level of mitochondria. For lipids to enter the mitochondria, they need to be complexed with carnitine derivatives. Primary or secondary carnitine deficiencies as well as deficiencies of enzymes involved in the metabolic pathway can interfere with this process, resulting in a myopathic picture.

Workup

Muscle enzyme levels may be elevated in the serum, but myoglobinuria is an unusual finding. Histochemical analysis of biopsied muscle shows excessive glycogen stores. Partial acid maltase deficiency may result in the adult form of Pompe disease and may be clinically indistinguishable from muscle phosphofructokinase deficiency (reported in at least one case) and glycogen debrancher enzyme deficiency. However, patients with glycogen debrancher enzyme deficiency eventually develop physical, electrical, or radiographic evidence of cardiac ventricle dilatation. Therefore, clinicians are advised to monitor for these changes by ECGs and examinations on an annual basis.

Treatment

Defects of acid maltase in their milder forms are discussed in Mental retardation/developmental delay with features that suggest storage disorders and in the article Lysosomal Storage Disease (ie, Pompe disease). The FDA has recently designated alglucosidase alfa (Myozyme) an orphan drug for use in infants with these disorders. The glycogen debrancher enzyme is discussed above in Mental retardation/developmental delay with features that suggest storage disorders, and muscle phosphofructokinase is discussed above in Cramps and weakness, often with some episodes of muscle breakdown and myoglobinuria, during or after heavy exercise.

FUTURE IN TREATMENT

In many instances, inborn errors of metabolism are difficult or impossible to treat. Novel therapeutic modalities, including gene therapy, need to be explored. Among available gene delivery systems, recombinant adeno-associated virus (AAV) shows promise for the treatment of metabolic disease. Despite the relatively low immunogenicity of AAV vectors, immune responses are also emerging as a factor requiring special attention as efforts accelerate towards human clinical translation. Trials targeting lipoprotein lipase deficiency are showing early evidence of efficacy.12

FERTILITY CONCERNS

Increasing numbers of individuals with inherited metabolic disorders are surviving into adulthood and considering their reproductive options. Lee discusses a practical approach to supporting such individuals, focusing on issues concerning fertility, the impact of pregnancy on metabolism and the metabolic disorder itself on the pregnancy, as well as highlighting the need to pay special attention during the postpartum period.13 He further opines that apart from pregnancies in women with phenylketonuria, data are lacking in this area. Collection of information within registries is needed to aid our understanding of potential problems and counseling of women and their partners.

CARBOHYDRATE METABOLISM DISORDERS

GLYCOGEN STORAGE DISEASES

Glycogen storage diseases are caused by deficiencies of enzymes involved in glycogen synthesis or breakdown; the deficiencies may occur in the liver or muscles and cause hypoglycemia or deposition of abnormal amounts or types of glycogen (or its intermediate metabolites) in tissues.

Inheritance for glycogen storage diseases (GSDs) is autosomal recessive except for GSD type VIII/IX, which is X-linked. Incidence is estimated at about 1/25,000 births, which may be an underestimate because milder subclinical forms may be undiagnosed. For a more complete listing of glycogen storage diseases, see Table Glycogen Storage Diseases and Disorders of Gluconeogenesis.

Age of onset, clinical manifestations, and severity vary by type, but symptoms and signs are most commonly those of hypoglycemia and myopathy. Diagnosis is suspected by history, examination, and detection of glycogen and intermediate metabolites in tissues by MRI or biopsy.

Defects in glycolysis (rare) may cause syndromes similar to GSDs. Deficiencies of phosphoglycerate kinase, phosphoglycerate mutase, and lactate dehydrogenase mimic the myopathies of GSD types V and VII; deficiencies of glucose transport protein 2 (Fanconi-Bickel syndrome) mimic the hepatopathy of other GSD types (eg, I, III, IV, VI).

GALACTOSEMIA

Galactosemia is caused by inherited deficiencies in enzymes that convert galactose to glucose. Symptoms and signs include hepatic and renal dysfunction, cognitive deficits, cataracts, and premature ovarian failure. Diagnosis is by enzyme analysis of RBCs.

Treatment is dietary elimination of galactose. Physical prognosis is good with treatment, but cognitive and performance parametres are often subnormal.

Galactose is found in dairy products, fruits, and

vegetables; autosomal recessive enzyme deficiencies cause 3 clinical syndromes.

Galactose-1-phosphate Uridyl Transferase Deficiency

This deficiency causes classic galactosemia. Incidence is 1/62,000 births; carrier frequency is 1/125. Infants become anorectic and jaundiced within a few days or weeks of consuming breast milk or lactose-containing formula. Vomiting, hepatomegaly, poor growth, lethargy, diarrhea, and septicemia (usually Escherichia coli) develop, as does renal dysfunction (eg, proteinuria, aminoaciduria, Fanconi syndrome), leading to metabolic acidosis and edema. Hemolytic anemia may also occur. Without treatment, children remain short and develop cognitive, speech, gait, and balance deficits in their teenage years; many also have cataracts, osteomalacia (caused by hypercalciuria), and premature ovarian failure. Patients with the Duarte variant have a much milder phenotype.

Galactokinase Deficiency

Patients develop cataracts from production of galactitol, which osmotically damages lens Fibres; idiopathic intracranial hypertension (pseudotumor cerebri) is rare. Incidence is 1/40,000 births.

Uridine Diphosphate Galactose 4-epimerase Deficiency

There are benign and severe phenotypes. Incidence of the benign form is 1/23,000 births in Japan; no incidence data are available for the more severe form. The benign form is restricted to RBCs and WBCs and causes no clinical abnormalities. The severe form causes a syndrome indistinguishable from classic galactosemia, although sometimes with hearing loss.

DIAGNOSIS AND TREATMENT

Diagnosis is suggested clinically and supported by

elevated galactose levels and the presence of reducing substances other than glucose (eg, galactose, galactose 1-phosphate) in the urine; it is confirmed by enzyme analysis of RBCs, hepatic tissue, or both. Most states require that neonates be screened for galactose-1-phosphate uridyl transferase deficiency.

Treatment is elimination of all sources of galactose in the diet, most notably lactose, which is a source of galactose present in all dairy products, including milk-based infant formulas and a sweetener used in many foods. A lactose-free diet prevents acute toxicity and reverses some manifestations (eg, cataracts) but may not prevent neurocognitive deficits. Many patients require supplemental Ca and vitamins. For patients with epimerase deficiency, some galactose intake is critical to ensure a supply of uridine-5?-diphosphate-galactose (UDP-galactose) for various metabolic processes.

DISORDERS OF FRUCTOSE METABOLISM

Deficiency of enzymes that metabolize fructose may be asymptomatic or cause hypoglycemia.

Fructose is a monosaccharide that is present in high concentrations in fruit and honey and is a constituent of sucrose and sorbitol.

Fructose 1-phosphate Aldolase Deficiency

This deficiency causes the clinical syndrome of hereditary fructose intolerance. Inheritance is autosomal recessive; incidence is estimated at 1/20,000 births. Infants are healthy until they ingest fructose; fructose 1-phosphate then accumulates, causing hypoglycemia, nausea and vomiting, abdominal pain, sweating, tremors, confusion, lethargy, seizures, and coma. Prolonged ingestion may cause cirrhosis, mental deterioration, and proximal renal tubular acidosis with urinary loss of phosphate and glucose.

Diagnosis is suggested by symptoms in relation to recent fructose intake and is confirmed by enzyme analysis of liver biopsy tissue or by induction of hypoglycemia by fructose infusion, 200 mg/kg IV. Diagnosis and identification of heterozygous carriers of the mutated gene can also be made by direct DNA analysis. Short-term treatment is glucose for hypoglycemia; long-term treatment is exclusion of dietary fructose, sucrose, and sorbitol. Many patients develop a natural aversion to fructose-containing food. Prognosis is excellent with treatment.

Fructokinase Deficiency

This deficiency causes benign elevation of blood and urine fructose levels (benign fructosuria). Inheritance is autosomal recessive; incidence is about 1/130,000 births. The condition is asymptomatic and diagnosed accidentally when a non-glucose reducing substance is detected in urine.

Deficiency of Fructose-1,6-bisphosphatase

This deficiency compromises gluconeogenesis and results in fasting hypoglycemia, ketosis, and acidosis. This deficiency can be fatal in neonates. Inheritance is autosomal recessive; incidence is unknown. Febrile illness can trigger episodes. Acute treatment is oral or IV glucose. Tolerance to fasting generally improves with age.

DISORDERS OF PYRUVATE METABOLISM

Inability to metabolize pyruvate causes lactic acidosis and a variety of CNS abnormalities. Pyruvate is an important substrate in carbohydrate metabolism.

Pyruvate Dehydrogenase Deficiency

Pyruvate dehydrogenase is a multi-enzyme complex responsible for the generation of acetyl CoA from pyruvate for the Krebs cycle. Deficiency results in elevation of pyruvate and thus elevation of lactic acid levels. Inheritance is X-linked or autosomal recessive.

Clinical manifestations vary in severity but include lactic acidosis and CNS malformations and other postnatal changes, including cystic lesions of the cerebral cortex, brain stem, and basal ganglia; ataxia; and psychomotor retardation. Diagnosis is confirmed by enzyme analysis of skin fibroblasts, DNA testing, or both. There is no clearly effective treatment, although a low-carbohydrate or ketogenic diet and dietary thiamin supplementation have been beneficial for some patients.

Pyruvate Carboxylase Deficiency

Pyruvate carboxylase is an enzyme important for gluconeogenesis from pyruvate and alanine generated in muscle. Deficiency may be primary, or secondary to deficiency of holocarboxylase synthetase, biotin, or biotinidase; inheritance for both is autosomal recessive and both result in lactic acidosis.

Primary deficiency incidence is < 1/250,000 births but may be higher in certain American Indian populations. Psychomotor retardation with seizures and spasticity are the major clinical manifestations. Labouratory abnormalities include hyperammonemia; lactic acidosis; ketoacidosis; elevated levels of plasma lysine, citrulline, alanine, and proline; and increased excretion of ?-ketoglutarate. Diagnosis is confirmed by enzyme analysis of cultured skin fibroblasts.

Secondary deficiency is clinically similar, with failure to thrive, seizures, and other organic aciduria. There is no effective treatment, but selected patients with primary deficiency and all those with secondary deficiencies should be given biotin supplementation 5 to 20 mg po once/day.

OTHER DISORDERS

Phosphoenolpyruvate carboxykinase deficiency impairs gluconeogenesis and results in symptoms and signs similar to the hepatic forms of GSD but without hepatic glycogen accumulation.

Other deficiencies include those of glycolytic enzymes or enzymes in the pentose phosphate pathway. Common examples are pyruvate kinase deficiency (Anemias Caused by Hemolysis: Embden-Meyerhof Pathway Defects) and glucose-6-phosphate dehydrogenase (G6PD) deficiency (see Anemias

Caused by Hemolysis: Glucose-6-Phosphate Dehydrogenase (G6PD) Deficiency), both of which may result in hemolytic anemia. Wernicke-Korsakoff syndrome (Vitamin Deficiency, Dependency, and Toxicity: Symptoms and Signs) is caused by a partial deficiency of transketolase, which is an enzyme for the pentose phosphate pathway that requires thiamin as a cofactor.

Chapter 13

Fats and Lipids

FAT (LIPIDS)

Despite some recent literature wrongly espousing the benefits of high fat intakes (i.e., intakes of 30 Per cent or more of total calories from fat), fat is a highly concentrated fuel that does nothing to improve athletic performance, body composition, or weight when taken in excess. The adult AMDR for total fat intake is 20 to 35 Per cent of total calories, and there is no scientific information suggesting that more than 25 Per cent of total calories from fat is generally better for athletes. However, for athletes who may have difficulty sustaining weight because of a massive energy expenditure (such as cross-country skiers) or who must sustain high weights (such as linemen on football teams), higher fat intakes (up to the AMDR limit of 35 Per cent) may be necessary.

Few Americans consume less than 35 Per cent of total calories from fat, so consumption of less fat is not easy, and unless steps are taken to provide sufficient energy from other substrates (mainly from more complex carbohydrates) to replace the eliminated fat, athletes may place themselves in an energy-deficit state that is, in itself, a detriment to performance. Therefore, while a reduction in fat intake is generally useful, a conscious effort should be made to provide enough total energy when fat intake is reduced. Since fat is more than twice as concentrated in calories than either protein or carbohydrate (9 calories per gram versus 4 calories per gram), more than twice as much food must be consumed to make up the difference in reduced fat.

Cholesterol, oils, butter, and margarine are all fats, or lipids, but each has slightly different characteristics. The one common attribute shared by lipids is that they are soluble in organic solvents but not soluble in water. (Anyone who has tried to mix Italian dressing knows this to be true. The oil in the dressing eventually rises to the top, no matter how hard the bottle is shaken.) The term fat is usually applied to lipids that are solid at room temperature, and the term oil is applied to lipids that are liquid at room temperature.

The most commonly consumed form of lipid is triglyceride, which consists of three fatty acids and one glycerol molecule (thus the name tri glyceride). Despite the numerous forms of lipids, we can obtain them all from the food supply, and we are also capable of making many types of lipids by combining carbon units from other substances. Nearly every cell in the body has the capacity to make cholesterol, which is why a person can have a high blood cholesterol leveleven when on a low-cholesterol diet.

We can also manufacture phospholipids, triglycerides, and oils. In fact, it is this ability to effectively manufacture different types of lipids that limits the necessity to consume large amounts of lipids.

STRUCTURE AND FUNCTION

FAT FUNCTIONS

A certain amount of fat, between 20 and 35 Per cent of total consumed calories, is necessary to ensure a sufficient energy and nutrient intake.

The fat-soluble vitamins-vitamins A,D,E, and K-must be delivered in a fat package. The essential fatty acids, which are needed for specific body functions but that we are incapable of synthesizing, must also be consumed. Some dietary fat is also needed to give us a feeling of satiety during the meal, creating the important physiological signal that it is time to stop eating. Dietary fats have a longer gastric emptying time than do carbohydrates, which contributes to the feeling of satiety. Of course, fat also makes foods taste good.

LIPIDS

All Lipids are hydrophobic: that's the one property they have in common. This group of molecules includes fats and oils, waxes, phospholipids, steroids (like cholesterol), and some other related compounds.

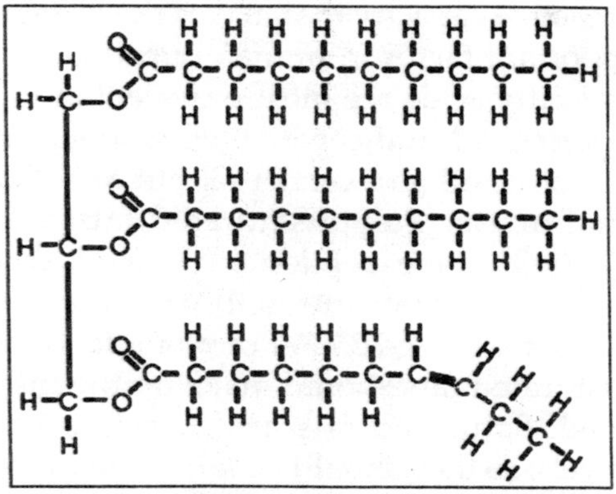

Fats and oils are made from two kinds of molecules: glycerol (a type of alcohol with a hydroxyl group on each of its three carbons) and three fatty acids joined by dehydration synthesis. Since there are three fatty acids attached, these are known as triglycerides."Bread" and pastries from a"bread factory" often contain mono- and diglycerides as"dough conditioners." Can you figure out what these molecules would look like? The main distinction between fats and oils is whether they're solid or liquid at room temperature, and this, as we'll soon see, is based on differences in the structures of the fatty acids they contain.

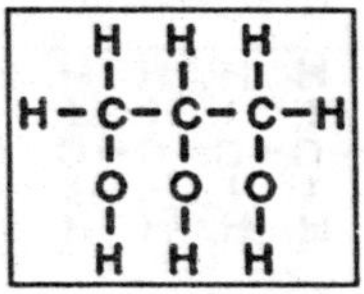

Lipid Structure

Lipids have different levels of saturation, a term that refers to the number of double bonds in the carbon chain. Fatty acids with no double bonds are saturated, those with one double bond are monounsaturated, and those with more than one double bond are polyunsaturated. Single bonds are stronger and less chemically reactive than double bonds, so the greater the number of double bonds, the greater the opportunity for the fatty acid to react with its chemical environment. It is this differential reactive capacity that makes the number of double bonds an important factor in human nutrition.

Saturated fatty acids are most prevalent in fats of animal origin, palm kernel oil, and coconut oil. Monounsaturated fats are highest in olive oil and canola oil but are also present in fats of animal origin. Polyunsaturated fats are highest in vegetable oils (with the exception of olive oil, which is more than 75 Per cent monounsaturated). In the context of a fat intake that does not exceed 35 Per cent of total calories, mono- and polyunsaturated fats should make up the majority of the fats consumed. Saturated fats are associated with higher cholesterol levels, so they should be minimized when possible. This is most easily achieved by reducing the consumption of animal fats, chocolate candies (often high in saturated tropical oils), fried foods, and high-fat dairy products.

STRUCTURE OF FATTY ACIDS

The"tail" of a fatty acid is a long hydrocarbon chain, making it hydrophobic. The"head" of the molecule is a carboxyl group which is hydrophilic. Fatty acids are the main component of soap, where their tails are soluble in oily dirt and their heads are soluble in water to emulsify and wash away the oily dirt. However, when the head end is attached to glycerol to form a fat, that whole molecule is hydrophobic.

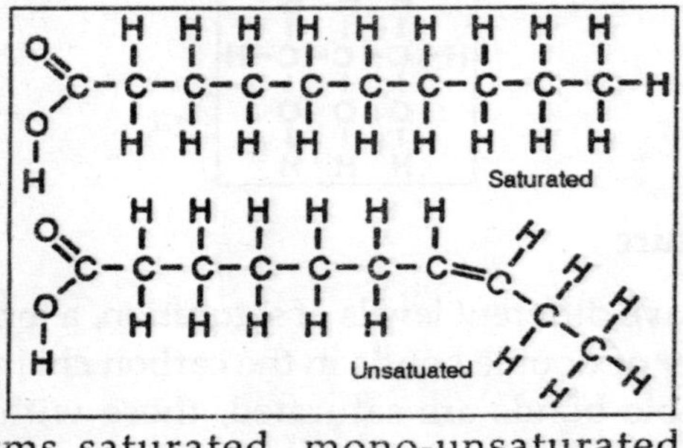

Saturated

Unsatuated

The terms saturated, mono-unsaturated, and polyunsaturated refer to the number of hydrogens attached to the hydrocarbon tails of the fatty acids as compared to the number of double bonds between carbon atoms in the tail. Fats, which are mostly from animal sources, have all single bonds between the carbons in their fatty acid tails, thus all the carbons are also bonded to the maximum number of hydrogens possible. Since the fatty acids in these triglycerides contain the maximum possible amouunt of hydrogens, these would be calledsaturated fats.

The hydrocarbon chains in these fatty acids are, thus, fairly straight and can pack closely together, making these fats solid at room temperature. Oils, mostly from plant sources, have some double bonds between some of the carbons in the hydrocarbon tail, causing bends or"kinks" in the shape of the molecules. Because some of the carbons share double bonds, they're not bonded to as many hydrogens as they could if they weren't double bonded to each other. Therefore these oils are called unsaturated fats.

Because of the kinks in the hydrocarbon tails, unsaturated fats can't pack as closely together, making them liquid at room temperature. Many people have heard that the unsaturated fats are"healthier" than the saturated ones. Hydrogenated vegetable oil (as in shortening and commercial peanut butters where a solid consistency is sought) started out as"good" unsaturated oil. However, this commercial product has had all the double bonds artificially broken and hydrogens artificially added (in a chemistry lab-type setting) to turn it into saturated fat that bears no resemblance to the original oil from which it came (so it will be solid at room temperature).

Trans

Cis

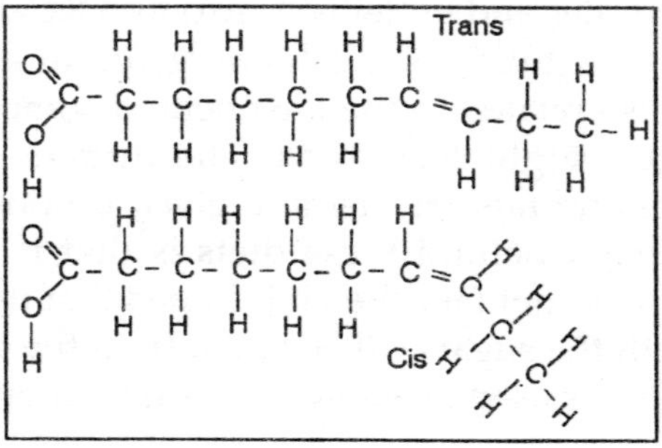

In unsaturated fatty acids, there are two ways the pieces of the hydrocarbon tail can be arranged around a C=C double bond. In cis bonds, the two pieces of the carbon chain on either side of the double bond are either both"up" or both"down," such that both are on the same side of the molecule. In trans bonds, the two pieces of the molecule are on opposite sides of the double bond, that is, one"up" and one"down" across from each other.

Naturally-occurring unsaturated vegetable oils have almost all cis bonds, but using oil for frying causes some of the cis bonds to convert to trans bonds. If oil is used only once like when you fry an egg, only a few of the bonds do this so it's not too bad. However, if oil is constantly reused, like in fast food French fry machines, more and more of the cis bonds are changed to trans until significant numbers of fatty acids with trans bonds build up. The reason this is of concern is that fatty acids with trans bonds are carcinogenic, or cancer-causing.

The levels of trans fatty acids in highly-processed, lipid-containing products such as margarine are quite high, and I have heard that the government is considering requiring that the amounts of trans fatty acids in such products be listed on the labels.

We need fats in our bodies and in our diet. Animals in general use fat for energy storage because fat stores 9 KCal/g of energy. Plants, which don't move around, can afford to store food for energy in a less compact but more easily accessible form, so they use starch (a carbohydrate, NOT A LIPID) for energy storage.

Carbohydrates and proteins store only 4 KCal/g of energy, so fat stores over twice as much energy/gram as fat. By the way, this is also related to the idea behind some of the high-carbohydrate weight loss diets. The human body burns carbohydrates and fats for fuel in a given proportion to each other. The theory behind these diets is that if they supply carbohydrates but not fats, then it is hoped that the fat needed to balance with the sugar will be taken from the dieter's body stores. Fat is also is used in our bodies to a) cushion vital organs like the kidneys and b) serve as insulation, especially just beneath the skin.

PHOSPHOLIPIDS

Phospholipids are made from glycerol, two fatty acids, and (in place of the third fatty acid) a phosphate group with some other molecule attached to its other end. The hydrocarbon tails of the fatty acids are still hydrophobic, but the phosphate group end of the molecule is hydrophilic because of the oxygens with all of their pairs of unshared electrons. This means that phospholipids are soluble in both water and oil.

An emulsifying agent is a substance which is soluble in both oil and water, thus enabling the two to mix. A"famous" phospholipid is lecithin which is found in egg yolk and soybeans. Egg yolk is mostly water but has a lot of lipids, especially cholesterol, which are needed by the developing chick. Lecithin is used to emulsify the lipids and hold them in the water as an emulsion. Lecithin is the basis of the classic emulsion known as mayonnaise.

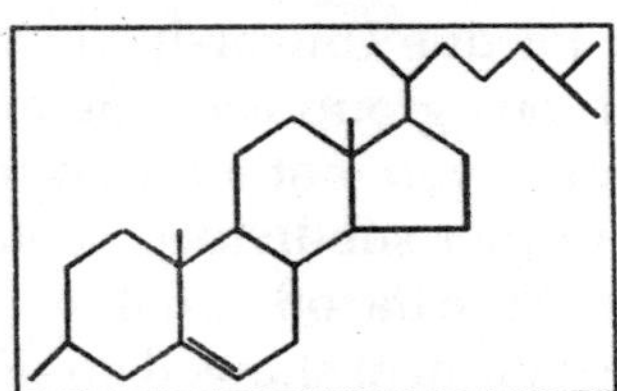

Our cell membranes are made mostly of phospholipids arranged in a double layer with the tails from both layers"inside" (facing towards each other) and the heads facing"out" (towards the watery environment) on both surfaces.

STEROIDS

The general structure of cholesterol consists of two six-membered rings side-by-side and sharing one side in common, a third six-membered ring off the top corner of the right ring, and a five-membered ring attached to the right side of that. The central core of this molecule, consisting of four fused rings, is shared by all steroids, including estrogen (estradiol), progesterone, corticosteroids such as cortisol (cortisone), aldosterone, testosterone, and Vitamin D. In the various types of steroids, various other groups/molecules are attached around the edges. Know how to draw the four rings that make up the central structure.

Cholesterol is not a"bad guy!" Our bodies make about 2 g of cholesterol per day, and that makes up about 85% of blood cholesterol, while only about 15% comes from dietary sources. Cholesterol is the precursor to our sex hormones and Vitamin D. Vitamin D is formed by the action of UV light in sunlight on cholesterol molecules that have"risen" to near the surface of the skin. At least one source I read suggested that people not shower immediately after being in the sun, but wait at least ½ hour for the new Vitamin D to be absorbed deeper into the skin. Our cell membranes contain a lot of cholesterol (in between the phospholipids) to help keep them"fluid" even when our cells are exposed to cooler temperatures.

Many people have hear the claims that egg yolk contains too much cholesterol, thus should not be eaten. An interesting study was done at Purdue University a number of years ago to test this. Men in one group each ate an egg a day, while men in another group were not allowed to eat eggs. Each of these groups was further subdivided such that half the men got"lots" of exercise while the other half were"couch potatoes." The results of this experiment showed no significant difference in blood cholesterol levels between egg-eaters and non-egg-eaters while there was a very significant difference between the men who got exercise and those who didn't.

Lipoproteins are clusters of proteins and lipids all tangled up together. These act as a means of carrying lipids, including cholesterol, around in our blood. There are two main categories of lipoproteins distinguished by how compact/dense they are. LDL or low density lipoprotein is the"bad guy," being associated with deposition of"cholesterol" on the walls of someone's arteries. HDL or high density lipoprotein is the"good guy," being associated with carrying"cholesterol" out of the blood system, and is more dense/more compact than LDL.

Triglycerides

The majority of consumed lipids are triglycerides, which contain three fatty acids and a glycerol molecule. Fat is stored in the form of triglycerides, which we manufacture when excess energy is consumed. We store triglycerides in adipose tissues (groups of fat cells) and inside muscle cells (intramuscular tri glyceride), both of which are available as an energy source when needed. When fat is burned as a source of energy, the stored triglycerides are taken out of storage, and each molecule is cleaved into its component fatty acids and glycerol molecule. Each fatty acid can then be broken apart (two carbon units at a time) and thrown into the cellular furnaces for the creation of ATP to form heat and provide the energy for muscular work. This process is referred to as the beta-oxidative metabolic pathway because burning fat, besides requiring some carbohydrate for its complete oxidation, also requires oxygen.

Glycerol is a unique lipid that is burned like a carbohydrate rather than a fat and is also an effective humectant (it holds water). Some long-endurance athletes find that adding glycerol to water helps them retain more water (i.e., to superhydrate) than if they consumed water alone. In an extremely hot and humid environment, water loss is likely to be higher than the athlete's fluid-replacement capacity, so beginning a competition in a superhydrated state may provide some advantages.

In studies of tennis and Olympic distance triathlon, athletes consuming glycerol in water prior to competition experienced some protective hyperhydration benefits when exercising in high heat.17,18 However, being superhydrated is likely to impart a degree of discomfort that requires some adaptation. Athletes who consume fluids containing glycerol often describe the feeling of holding extra body water as making them feel"like a water bag," or"heavy," or"stiff." Still, they maintain that how they feel at the end of a race is more important than how they feel at the beginning, so adding glycerol to precompetition fluids has become a standard protocol for some athletes.

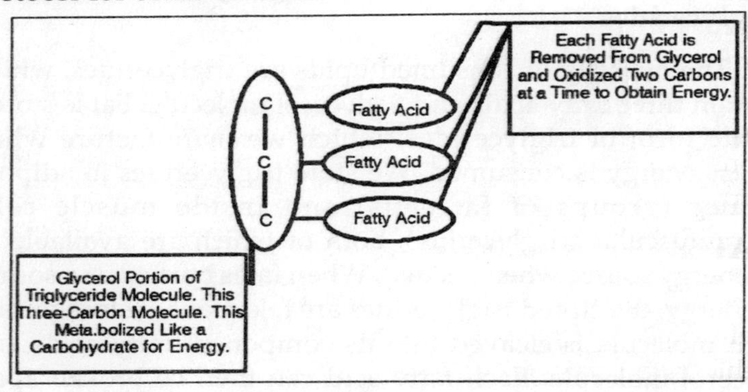

FATTY ACIDS

Linoleic (omega-6) and linolenic (omega-3) fatty acids are the essential fatty acids; although they are needed for metabolic processes, we are incapable of synthesizing them. The"omega-6" classification means that these fatty acids, which are polyunsaturated, have the last double bond six carbons from the end of the carbon chain.

The"omega-3" classification means that these fatty acids have the last double bond three carbons from the end of the carbon chain. Linoleic acid is an essential part of lipid membranes and is required for normal skin health. Linolenic acid is necessary for neural function and growth. The AMDR for omega-6 andomega-3 fatty acids is 5 to 10 and.6 to 1.2 grams per day, respectively.11Both fatty acids are easily obtained from vegetable oils (corn, safflower, canola, and so on) and the oils of cold-water fish.

Fish liver oils that are high in omega-3 acids have recently received much attention. These oils have been shown to reduce the ability of red blood cells to congregate, thereby decreasing the chance for an unwanted blood clot to form. This reduces the risk of a heart attack, which is most commonly caused by a clot formation in one of the major heart arteries. The oils from cold-water fish are the main source of the omega-3 fatty acids eicosapentanoic acid (EPA) and docosohexanoic acid (DHA).

Even a once-weekly consumption of cold-water fish (salmon, albacore tuna, Atlantic herring) is sufficient to significantly reduce the risk of heart attack and stroke.11Despite these findings, excessive intake of these fish oils may cause problems, including an increase in cellular oxidative damage. The best rule of thumb is to make fish consumption a regular part of the weekly diet so that supplemental intake of omega-3 fatty acids is unnecessary.

FAT REQUIREMENTS

From an exercise standpoint, there is little reason to believe that increasing fat consumption results in improving athletic performance, unless the increase in fat intake is the only reasonable means for the athlete to obtain sufficient energy. For the athlete who needs more than 4,000 calories each day to meet the combined demands of growth, exercise, and maintenance, moderate increases in dietary fat (preferably from plant and fish sources) may be needed. Since fat is a more concentrated form of energy than either carbohydrate or protein, more energy can be consumed in a smaller food package if the foods contain more fat.

LIPIDS AND PHYSICAL ACTIVITY

Even the leanest, healthiest athletes have a substantial energy pool of stored lipids. The average storage in adipose tissue ranges between 50,000 and 100,000 calories, or enough energy to walk or run 500 to 1,000 miles19 (800 to 1,600 kilometres) without a refueling stop. In addition, athletes store approximately 2,000 to 3,000 calories of lipids inside the muscle tissue.20 These lipids, which are stored in the form of triglycerides, are available as a fuel under the proper conditions of oxygen availability. Maximal fat oxidation occurs at 60 to 65 Per cent VO_2 max, but at higher levels of VO2max there is insufficient oxygen to derive the majority of energy used from fat catabolism.

Triglycerides stored in adipose tissue are cleaved into their component molecules of glycerol and fatty acids and transported to the blood plasma. The glycerol is available to all tissues for energy metabolism, and the free fatty acids are transported to working muscles where they are oxidized for energy.

The triglycerides stored in working muscles are cleaved into glycerol and fatty acids, and the fatty acids can be oxidized for energy where they are resident. The glycerol can also be burned for energy in the working muscle or can be transported to the blood plasma as a source of energy for other tissues.

The lower the exercise intensity, the greater the proportion of fat burned to satisfy energy needs. As exercise intensity increases, the proportion of fat burned decreases and the proportion of carbohydrate burned increases. It is this basic reality that is behind why so many people do low-intensity activity to burn fat and lower body fat levels. However, the proportion of fat burned should not be confused by the total amount of fat burned at different intensities of physical activity.

As exercise intensity increases, the total number of calories burned per unit of time also increases. Although there may be a decrease in the proportion of fat burned to satisfy total energy needs in higher-intensity activity, the total volume of fat burned is greater because the total energy requirement is higher.

The take-away message from this metabolic reality is that athletes interested in lowering body fat should exercise at least as high as 65 Per cent of VO2 max for the duration of their workouts to optimize the total mass of fat that is burned. Exercising at lower intensities burns a greater proportion of fat but less total fat than exercising at higher intensities.

BIOSYNTHESIS OF FATTY ACID

Identification of Genes in Biosynthesis of Fatty Acids and Cholesterol Pathways Using Microarrays Sterol regulatory binding proteins (SREBP) are a family of basic helix-loop-helix leucine zipper transcription factors that regulate biosynthesis of fatty acids and cholesterol. To identify candidate genes in the fatty acid synthesis pathway, transgenic mice over expressing SREBP-1a and SREBP-2 isofroms in liver were used. Labeled cRNA was prepared from liver and hybridized to mouse expression microarrays interrogating ~30,000 genes/ ESTs. Some of the known genes, ACC, FAS, and ATP citrate lyase regulated by over expression of SREBP were identified by microarray analysis.

A novel gene, long chain fatty acyl elongase (LCE) regulated by the SREBP1a and 2 was also identified. LCE was over expressed with a fold change of 25 and 9 in SREBP-1a and SREBP-2 mice respectively. The confirmation of LCE was done by northern blot analysis, sequence homology searches and biochemical tests. This enzyme specifically catalyzes the elongation of 12, 14 and 16 carbons but does not elongate the C-18 fatty acids or very long chain fatty acids.

DEGRADATION OF FATTY ACIDS

Multi-million pounds of viscera are generated from the catfish filleting process. Our previous study showed that catfish viscera contained 30-35% crude fat, and catfish visceral oil contained health-promoting fatty acids. Catfish visceral oils show potential for use as nutrient supplement and as cooking oil. Information on thermal stability and degradation of catfish visceral oils is not available.

The objective of this study was to determine weight loss (%) of catfish visceral oils and fatty acids after having been subjected to heat treatment. Oil samples included crude, degummed, neutralized, bleached, and deodorized oils.

The oils and individual fatty acids, FAs (C14:0, C16:0, C18:0, C20:0, C16:1, C18:1, C18:2, C18:3, C20:2, C20:4, C22:6) were subjected to the Thermogravimetric Analyser. Samples (0.5-1 mg) was placed in the furnace and heated up to 600°C at the rate of 5°C/min. Weight loss (%) of oils and fatty acids were calculated based on the original weight of non-heated samples. Duplicate experiments were conducted.

Over 90% weight loss of all FAs (except C18:1 and C20:2) was observed at 250°C. Less than 2% remaining weight of FAs (except C20:2) was observed at 350°C. Saturated FAs completely decomposed above 400°C. Comparing C18:1 with C18:2 and C18:3, weight loss increased with increased double bonds. Between 200-550°C, weight loss of oils slightly increased with increased temperature, regardless of the purification process. There was, however, no distinct weight loss of all oils below 300°C. Oils with less impurity had greater weight loss especially at 450-500°C. At 550°C, all oil samples completely decomposed.

Knowledge of thermal stability or degradation (weight loss through eVapouration or decomposition) of catfish oils during purification or cooking is essential for the design of processing units, analysis of production cost, and cooking quality if to be used as cooking or seasoning oil.

CHOLESTROL BIOSYNTHESIS AND DEGRADATION

CHOLESTEROL BIOSYNTHESIS AND CATABOLISM

The formation of cholesterol by the liver, regulated by a type of homeostatic control, is influenced by the hepatic level offree cholesterol or glycogen, or both, and by the availability of the proper enzymes.

Here the factors affecting the hepaticbiosynthesis of cholesterol are detailed, including the influence of dietary composition and fasting. The specific effects oncholesterol synthesis of feeding cholesterol, various steroids, saturated and unsaturated fats and thyroid hormone are reviewed. Also considered are the effects of radiation, glycolysis and various pharmacologic agents. In the degradation of cholesterol, the chief metabolic end products are bile acids, although cholesterol may also be converted into sex hormones and adrenocortical hormones. The steps involved in the degradation of cholesterol to bileacids are reviewed, along with the various factors which increase or decrease cholesterol catabolism. Since degradation of cholesterol depends on the levels of bile salts in the liver, any factor which will increase the turnover of bile acids will also accelerate cholesterol catabolism and may in turn lower serum cholesterol.

CHOLESTEROL DESCRIPTION

Cholesterol is a vital constituent of cell membranes and the precursor for steroid hormones and bile acids. It is clearly essential to life, yet its deposition in arteries is associated with cardiovascular disease and stroke, two leading causes of death in humans. In a healthy organism, an intricate balance is maintained between the biosynthesis, utilization, and transport of cholesterol, keeping its harmful deposition to a minimum."

HOLESTEROL A CRUCIAL COMPONENT OF CELL MEMBRANES

- Cholesterol is the third major lipid in membranes.
- Cholesterol is compact,rigid,hydrophobicmolecule with a polar hydroxyl group.
- The hydroxyl group of cholesterol is orientedtowards the aqueous phase (polar heads of phospholipids) and gives hydrophilic character to this end of cholesterol molecule.
- The cholesterol to polar lipid ratios affect stability, fluidity, permeability, and protein mobility. Cholesterol also controls the microstructure of membranes.
- Cholesterol is especially abundant in myelinated structures of the brain and central nervous system and almost absent from the inner mitochondrial membrane.

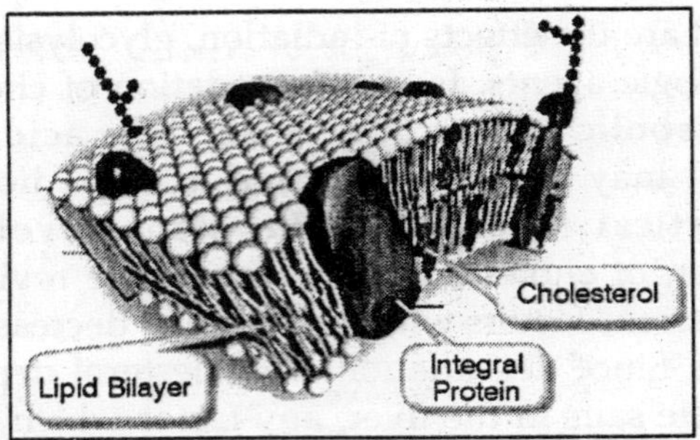

CHOLESTEROL IS PRECURSOR OF BILE SALTS

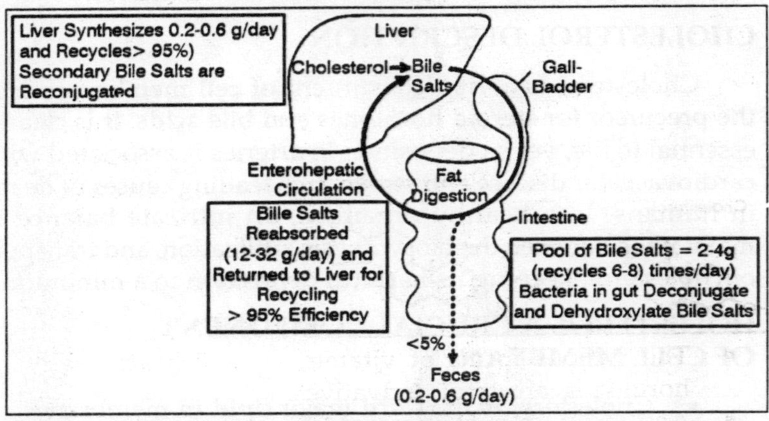

- Cholesterol is abundant in bile.
- The solubilization of hydrophobic molecules of cholesterol is aided by bile phospholipids and bile salts. They prevent cholesterol from precipitating in gallbladder in form of gallstones.
- Bile salts are metabolites of cholesterol.
- Cholesterol protects gallbladder membranes from irritating and harmful effects of bile acids and bile salts.

Cholesterol is the precursor of steroid hormones:

Testosterone: A Male Sex Hormone
Estradol: a Female Sex Hormone
Estrone: A Female Sex Hormone

Cholesterol is the precursor of ALL steroid hormones including progesterone, estrogen, testosterone,corticosteroids, and aldosterone. Thus, cholesterol is crucial for human reproduction, response to the environment, and preservation of human species.

CHOLESTEROL IS PRECURSOR OF VITAMIN D

- The various forms of vitamin D, which are really hormones, aresterol derivatives.
- Active vitamin D increases Ca2+concentration in serum by promoting the intestinal absorption of dietary Ca2+. This increases the deposition of Ca2+ in bones and teeth.

BIOSYNTHESIS OF CHOLESTEROL

Slightly less than half of the cholesterol in the body derives from biosynthesis de novo. Biosynthesis in the liver accounts for approximately 10%, and in the intestines approximately 15%, of the amount produced each day. Cholesterol synthesis occurs in the cytoplasm and microsomes from the two-carbon acetate group of acetyl-CoA.

The process has five major steps:

- Acetyl-CoAs are converted to 3-hydroxy-3-methylglutaryl-CoA (HMG-CoA).

- HMG-CoA is converted to mevalonate.
- Mevalonate is converted to the isoprene based molecule, isopentenyl pyrophosphate (IPP), with the concomitant loss of CO_2.
- IPP is converted to squalene.
- Squalene is converted to cholesterol.

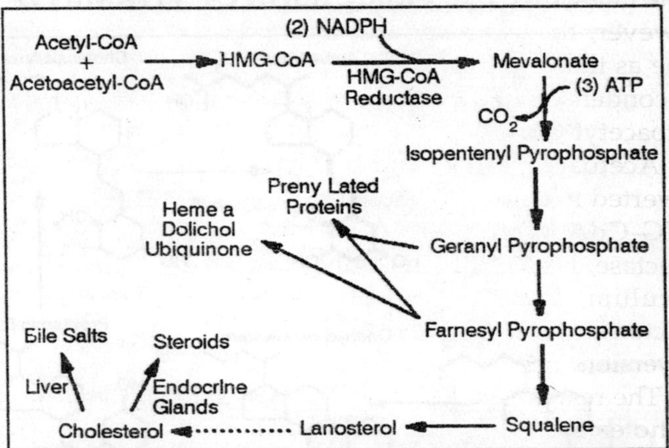

Pathway of cholesterol biosynthesis. Synthesis begins with the transport of acetyl-CoA ffrom the mitochondrion to the cytosol. The rate limiting step occurs at the 3-hydroxy-3-methlgltarlCoA(HMGCoA)redcataseHMGRcataledstepThe phosphorylation reactions are required to solubilize the isoprenoid intermediates in the pathway. Intermediates in the pathway are used for the synthesis of prenylated proteins, dolichol, coenzyme Q and the side chain of heme a.

The acetyl-CoA utilized for cholesterol biosynthesis is derived from an oxidation reaction (eg, fatty acids or pyruvate) in the mitochondria and is transported to the cytoplasm by the same process as that described for fatty acid synthesis. Acetyl-CoA can also be derived from cytoplasmic oxidation of ethanol by acetyl-CoA synthetase. All the reduction reactions of cholesterol biosynthesis use NADPH as a cofactor. The isoprenoid intermediates of cholesterol biosynthesis can be diverted to other synthesis reactions, such as those for dolichol (used in the synthesis of N-linked glycoproteins, coenzyme Q (of the oxidative phosphorylation) pathway or the side chain of heme a. Additionally, these intermediates are used in the lipid modification of some proteins.

Acetyl-CoA units are converted to mevalonate by a series of reactions that begins with the formation of HMG-CoA. Unlike the HMG-CoA formed during ketone body synthesis in the mitochondria, this form is synthesized in the cytoplasm. However, the pathway and the necessary enzymes are the same as those in the mitochondria. Two moles of acetyl-CoA are condensed in a reversal of the thiolase reaction, forming acetoacetyl-CoA.

Acetoacetyl-CoA and a third mole of acetyl-CoA are converted to HMG-CoA by the action of HMG-CoA synthase. HMG-CoA is converted to mevalonate by HMG-CoA reductase, HMGR (this enzyme is bound in the endoplasmic reticulum, ER). HMGR absolutely requires NADPH as a cofactor and two moles of NADPH are consumed during the conversion of HMG-CoA to mevalonate.

The reaction catalyzed by HMGR is the rate limiting step of cholesterol biosynthesis, and this enzyme is subject to complex regulatory controls.

Mevalonate is then activated by three successive phosphorylations, yielding 5-pyrophosphomevalonate. In addition to activating mevalonate, the phosphorylations maintain its solubility, since otherwise it is insoluble in water. After phosphorylation, an ATP-dependent decarboxylation yields iscpentenyl pyrophosphate, IPP, an activated isoprenoid molecule. Isopentenyl pyrophosphate is in equilibrium with its isomer, dimethylallyl pyrophosphate, DMPP.

One molecule of IPP condenses with one molecule of DMPP to generate geranyl pyrophosphate, GPP. GPP further condenses with another IPP molecule to yield farnesyl pyrophosphate, FPP. Finally, the NADPH-requiring enzyme, squalene synthase catalyzes the head-to-tail condensation of two molecules of FPP, yielding squalene (squalene synthase also is tightly associated with the endoplasmic reticulum). Squalene undergoes a two step cyclization to yield lanosterol.

The first reaction is catalyzed by squalene

monooxygenase. This enzyme uses NADPH as a cofactor to introduce molecular oxygen as an epoxide at the 2,3 position of squalene. Through a series of 19 additional reactions, lanosterol is converted to cholesterol.

REGULATING CHOLESTEROL SYNTHESIS

Normal healthy adults synthesize cholesterol at a rate of approximately 1g/day and consume approximately 0.3g/day. A relatively constant level of cholesterol in the body (150 - 200 mg/dL) is maintained primarily by controlling the level of de novo synthesis. The level of cholesterol synthesis is regulated in part by the dietary intake of cholesterol. Cholesterol from both diet and synthesis is utilized in the formation of membranes and in the synthesis of the steroid hormones and bile acids (see below). The greatest proportion of cholesterol is used in bile acid synthesis.

The cellular supply of cholesterol is maintained at a steady level by three distinct mechanisms:

- Regulation of HMGR activity and levels
- Regulation of excess intracellular free cholesterol through the activity of acyl-CoA:cholesterol acyltransferase, ACAT
- Regulation of plasma cholesterol levels via LDL receptor-mediated uptake and HDL-mediated reverse transport.

Regulation of HMGR activity is the primary means for controlling the level of cholesterol biosynthesis. The enzyme is controlled by four distinct mechanisms: feed-back inhibition, control of gene expression, rate of enzyme degradation and phosphorylation-dephosphorylation.

The first three control mechanisms are exerted by cholesterol itself. Cholesterol acts as a feed-back inhibitor of pre-existing HMGR as well as inducing rapid degradation of the enzyme. The latter is the result of cholesterol-induced polyubiquitination of HMGR and its degradation in the proteosome (see proteolytic degradation below). This ability of cholesterol is a consequence of the sterol sensing domain, SSD of HMGR. In addition, when cholesterol is in excess the amount of mRNA for HMGR is reduced as a result of decreased expression of the gene. The mechanism by which cholesterol (and other sterols) affect the transcription of the HMGR gene is described below under regulation of sterol content.

Regulation of HMGR through covalent modification occurs as a result of phosphorylation and dephosphorylation. The enzyme is most active in its unmodified form. Phosphorylation of the enzyme decreases its activity. HMGR is phosphorylated by AMP-activated protein kinase, AMPK (this is not the same as cAMP-dependent protein kinase, PKA). AMPRK itself is activated via phosphorylation. The phosphorylation of AMPK is catalyzed by AMPK kinase.

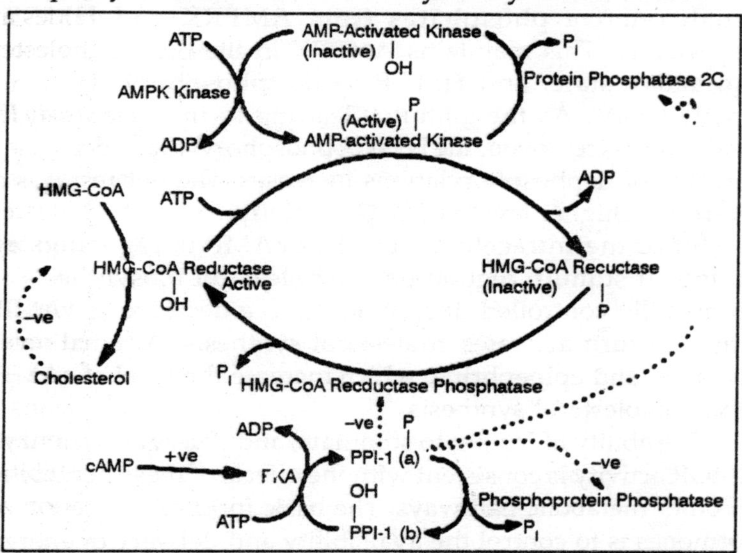

Regulation of HMGR by covalent modification. HMGR is most active in the dephosphorylated state. Phosphorylation is catalyzed by AMP-activated protein kinase, AMPK, (used to be termed HMGR kinase), an enzyme whose activity is also regulated by phosphorylation. Phosphorylation of AMPK is catalyzed by AMPK kinase. Hormones such as glucagon and epinephrine negatively affect cholesterol biosynthesis by increasing the activity of the inhibitor of phosphoprotein phosphatase inhibitor-1, PPI-1. Conversely, insulin stimulates the removal of phosphates and, thereby, activates HMGR activity. Additional regulation of HMGR occurs through an inhibition of its' activity as well as of its' synthesis by elevation in intracellular cholesterol levels. This latter phenomenon involves the transcription factor SREBP described. The activity of HMGR is additionally controlled by the cAMP signaling pathway. Increases in cAMP lead to activation of cAMP-dependent protein kinase, PKA. In the context of HMGR regulation, PKA phosphorylates phosphoprotein phosphatase inhibitor-1 (PPI-1) leading to an increase in its' activity. PPI-1 can inhibit the activity of numerous phosphatases including protein phosphatase 2C and HMG-CoA reductase phosphatase which remove phosphates from AMPRK and HMGR, respectively. This maintains AMPRK in the phosphorylated and active state, and HMGR in the phosphorylated and inactive state. As the stimulus leading to increased cAMP production is removed, the level of phosphorylations decreases and that of dephosphorylations increases. The net result is a return to a higher level of HMGR activity.

Since the intracellular level of cAMP is regulated by hormonal stimuli, regulation of cholesterol biosynthesis is hormonally controlled. Insulin leads to a decrease in cAMP, which in turn activates cholesterol synthesis. Alternatively, glucagon and epinephrine, which increase the level of cAMP, inhibit cholesterol synthesis.

The ability of insulin to stimulate, and glucagon to inhibit, HMGR activity is consistent with the effects of these hormones on other metabolic pathways. The basic function of these two hormones is to control the availability and delivery of energy to all cells of the body.

Long-term control of HMGR activity is exerted primarily through control over the synthesis and degradation of the enzyme. When levels of cholesterol are high, the level of expression of the HMGR gene is reduced. Conversely, reduced levels of cholesterol activate expression of the gene. Insulin also brings about long-term regulation of cholesterol metabolism by increasing the level of HMGR synthesis.

Proteolytic Regulation of HMG-CoA Reductase

The stability of HMGR is regulated as the rate of flux through the mevalonate synthesis pathway changes. When the flux is high the rate of HMGR degradation is also high. When the flux is low, degradation of HMGR decreases. This phenomenon can easily be observed in the presence of the statin drugs.

HMGR is localized to the ER and like SREBP (see below) contains a sterol-sensing domain, SSD. When sterol levels increase in cells there is an concommitant increase in the rate of HMGR degradation. The degradation of HMGR occurs within the proteosome, a multiprotein complex dedicated to protein degradation. The primary signal directing proteins to the proteosome is ubiquitination. Ubiquitin is a 7.6kDa protein that is covalently attached to proteins targeted for degradation by ubiqitin ligases. These enzymes attach multiple copies of ubiquitin allowing for recognition by the proteosome. HMGR has been shown to be ubiquitinated prior to its degradation. The primary sterol regulating HMGR degradation is cholesterol itself. As the levels of free cholesterol increase in cells, the rate of HMGR degradation increases.

UTILIZATION OF CHOLESTEROL

Cholesterol is transported in the plasma predominantly as cholesteryl esters associated with lipoproteins. Dietary cholesterol is transported from the small intestine to the liver within chylomicrons. Cholesterol synthesized by the liver, as well as any dietary cholesterol in the liver that exceeds hepatic needs, is transported in the serum within LDLs. The liver synthesizes VLDLs and these are converted to LDLs through the action of endothelial cell-associated lipoprotein lipase.

Cholesterol found in plasma membranes can be extracted by HDLs and esterified by the HDL-associated enzyme LCAT. The cholesterol acquired from peripheral tissues by HDLs can then be transferred to VLDLs and LDLs via the action of cholesteryl ester transfer protein (apo-D) which is associated with HDLs. Reverse cholesterol transport allows peripheral cholesterol to be returned to the liver in LDLs. Ultimately, cholesterol is excreted in the bile as free cholesterol or as bile salts following conversion to bile acids in the liver.

REGULATION OF CELLULAR STEROL CONTENT

The continual alteration of the intracellular sterol content occurs through the regulation of key sterol synthetic enzymes as well as by altering the levels of cell-surface LDL receptors. As cells need more sterol they will induce their synthesis and uptake, conversely when the need declines synthesis and uptake are decreased. Regulation of these events is brought about primarily by sterol-regulated transcription of key rate limiting enzymes and by the regulated degradation of HMGR. Activation of transcriptional control occurs through the regulated cleavage of the membrane-bound transcription factor sterol regulated element binding protein,

Sterol control of transcription affects more than 30 genes involved in the biosynthesis of cholesterol, triacylglycerols, phospholipids and fatty acids. Transcriptional control requires the presence of an octamer sequence in the gene termed the sterol regulatory element, SRE-1. It has been shown that SREBP is the transcription factor that binds to SRE-1 elements. It turns out that there are 2 distinct SREBP gene, SREBP-1 and SREBP-2. In addition, the SREBP-1 gene encodes 2 proteins, SREBP-1a and SREBP-1c as a consequence of alternative exon usage.

All 3 proteins are proteolytically regulated by sterols. Full-length SREBPs have several domains and are embedded in the membrane of the endoplasmic reticulum (ER). The N-terminal domain contains a transcription factor motif of the basic helix-loop-helix (bHLH) type that is exposed to the cytoplasmic side of the ER.

There are 2 transmembrane spanning domains followed by a large C-terminal domain also exposed to the cytosolic side. When sterols are scarce, cleavage of the full-length SREBP takes place with the result being that the N-terminal bHLH motif is released into the cytosol. The bHLH domain then migrates to the nucleus to direct transcription. Conversely, when sterols are abundant, cleavage of SREBP is inhibited. To control the level of SREBP-mediated transcription, the soluble bHLH domain is itself subject to rapid proteolysis.

The cleavage of SREBP is carried out by 2 distinct enzymes, one of which is regulated by sterols. The regulated cleavage occurs in the lumenal loop between the 2 transmembrane domains. This cleavage is catalyzed by site-1 protease, S1P. High sterol content blocks the activity of S1P. The second cleavage, catalyzed by site-2 protease, S2P, occurs in the first transmembrane span, leading to release of active SREBP.

Additional studies on sterol-regulated gene expression demonstrated that cleavage of SREBP by S1P is controlled by the level and action of an additional protein termed, SREBP cleavage-activating protein, SCAP. SCAP is a large protein also found in the ER membrane and contains at least 8 transmembrane spans. The C-terminal portion, which extends into the cytosol, has been shown to interact with the C-terminal domain of SREBP. This C-terminal region of SCAP contains 4 motifs called WD40 repeats.

The WD40 repeats are required for interaction of SCAP with SREBP. Interestingly, the N-terminus of SCAP, including membrane spans 2-6, resembles HMGR which itself is subject to sterol-stimulated degradation (see above). This shared motif is called the sterol sensing domain, SSD. Several proteins whose functions involve sterols also contain the SSD.

These include patched, an important development regulating receptor whose ligand, hedgehog, is modified by attachment of cholesterol and the Neimann Pick C1 (NPC1) protein which is involved in cholesterol transport in the secretory pathway. NPC1 is one of several genes whose activities, when disrupted, lead to severe neurological dysfunction. The function of SCAP is to positively stimulate S1P-mediated cleavage of SREBP. The function of sterols is to inhibit this positive action of SCAP. The activity of SCAP involves movement from the ER to the Golgi and back. Because the C-terminus of SCAP interacts with SREBP, movement of SCAP takes SREBP along for the ride. When sterols are low, SCAP and SREBP move to the Golgi. This transit is required for SREBP cleavage as S1P is Golgi-localized. When sterols are high, movement of SCAP is halted. Thus, the overall effect of sterols is to regulate the ability of SCAP to present SREBP to S1P.

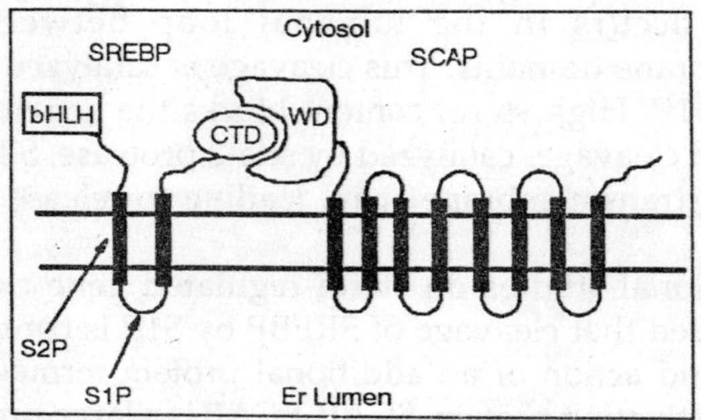

Diagramatic representation of the interactions between SREBP and SCAP in the membrane of the ER CTD = C-terminal domain

BILE ACIDS SYNTHESIS AND UTILIZATION

The end products of cholesterol utilization are the bile acids, synthesized in the liver. Synthesis of bile acids is one of the predominant mechanisms for the excretion of excess cholesterol. However, the excretion of cholesterol in the form of bile acids is insufficient to compensate for an excess dietary intake of cholesterol.

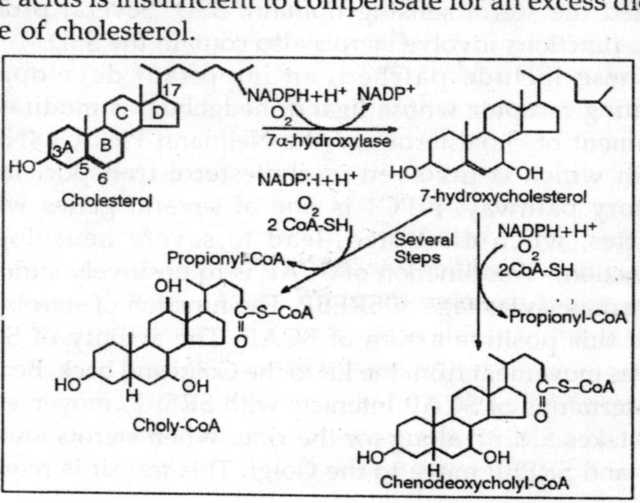

Synthesis of the 2 primary bile acids, cholic acid and chenodeoxycholic acid. The reaction catalyzed by the 7α-hydroxylase is the rate limiting step in bile acid synthesis. Conversion of 7α-hydroxycholesterol to the bile acids requires several steps not shown in detail in this image. Only the relevant co-factors needed for the synthesis steps are shown.

The most abundant bile acids in human bile are chenodeoxycholic acid (45%) and cholic acid (31%). These are referred to as the primary bile acids. Within the intestines the primary bile acids are acted upon by bacteria and converted to the secondary bile acids, identified as deoxycholate (from cholate) and lithocholate (from chenodeoxycholate). Both primary and secondary bile acids are reabsorbed by the intestines and delivered back to the liver via the portal circulation.

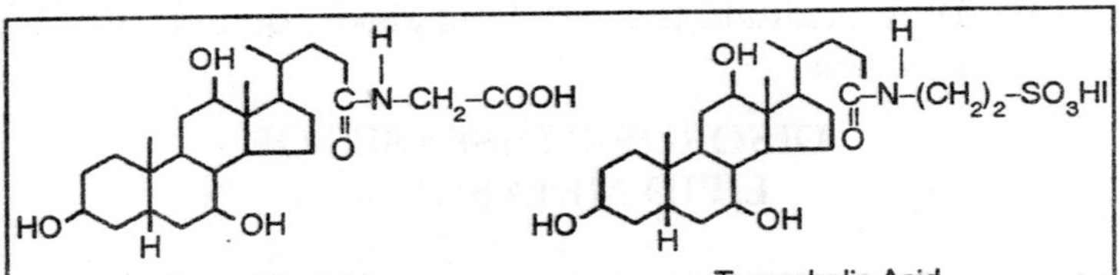

Glycochlic Acid   Taurocholic Acid

STRUCTURE OF CONJUGATED CHOLIC ACIDS

Within the liver the carboxyl group of primary and secondary bile acids is conjugated via an amide bond to either glycine or taurine before their being resecreted into the bile canaliculi.

These conjugation reactions yield glycoconjugates and tauroconjugates, respectively. The bile canaliculi join with the bile ductules, which then form the bile ducts. Bile acids are carried from the liver through these ducts to the gallbladder, where they are stored for future use.

The ultimate fate of bile acids is secretion into the intestine, where they aid in the emulsification of dietary lipids. In the gut the glycine and taurine residues are removed and the bile acids are either excreted (only a small Percentage) or reabsorbed by the gut and returned to the liver. This process of secretion from the liver to the gallbladder, to the intestines and finally reabsorbtion is termed the enterohepatic circulation.

CLINICAL SIGNIFICANCE OF BILE ACID SYNTHESIS

Bile acids perform four physiologically significant functions:

- Their synthesis and subsequent excretion in the feces represent the only significant mechanism for the elimination of excess cholesterol.
- Bile acids and phospholipids solubilize cholesterol in the bile, thereby preventing the precipitation of cholesterol in the gallbladder.
- They facilitate the digestion of dietary triacylglycerols by acting as emulsifying agents that render fats accessible to pancreatic lipases.
- They faciiitate the intestinal absorption of fat-soluble vitamins.

DISORDER/ DISEASES OF LIPID METABOLISM

DISORDERS OF LIPID METABOLISM

Fats (lipids) are an important source of energy for the body. The body's store of fat is constantly broken down and reassembled to balance the body's energy needs with the food available. Groups of specific enzymes help the body break down and process fats. Certain abnormalities in these enzymes can lead to the buildup of specific fatty substances that normally would have been broken down by the enzymes. Over time, accumulations of these substances can be harmful to many organs of the body. Disorders caused by the accumulation of lipids are called lipidoses. Other enzyme abnormalities prevent the body from converting fats into energy normally. These abnormalities are called fatty acid oxidation disorders.

GAUCHER'S DISEASE

Gaucher's disease is caused by a buildup of

glucocerebrosides in tissues. Children who have the infantile form usually die within a year, but children and adults who develop the disease later in life may survive for many years.

In Gaucher's disease, glucocerebrosides, which are a product of fat metabolism, accumulate in tissues. Gaucher's disease is the most common lipidosis. The disease is most common among Ashkenazi (Eastern European) Jews. Gaucher's disease leads to an enlarged liver and spleen and a brownish pigmentation of the skin. Accumulations of glucocerebrosides in the eyes cause yellow spots called pingueculae to appear. Accumulations in the bone marrow can cause pain and destroy bone.

RARE HEREDITARY DISORDERS OF LIPID METABOLISM

Wolman's disease results when specific types of cholesterol and glycerides accumulate in tissues. This disease causes enlargement of the spleen and liver. Calcium deposits in the adrenal glands cause them to harden, and fatty diarrhea (steatorrhea) also occurs. Infants with Wolman's disease usually die by 6 months of age.

Cerebrotendinous xanthomatosis occurs when cholestanol, a product of cholesterol metabolism, accumulates in tissues. This disease eventually leads to uncoordinated movements, dementia, cataracts, and fatty growths (xanthomas) on tendons. The disabling symptoms often appear after age 30. If started early, the drug chenodiol helps prevent progression of the disease, but it cannot undo any damage already done.

In sitosterolemia, fats from fruits and vegetables accumulate in blood and tissues. The buildup of fats leads to atherosclerosis, abnormal red blood cells, and xanthomas on tendons. Treatment consists of reducing the intake of foods that are rich in plant fats, such as vegetable oils, and takingcholestyramine resin.

In Refsum's disease, phytanic acid, which is a product of fat metabolism, accumulates in tissues. A buildup of phytanic acid leads to nerve and retinal damage, spastic movements, and changes in the bone and skin. Treatment involves avoiding eating green fruits and vegetables that contain chlorophyll. Plasmapheresis, in which phytanic acid is removed from the blood, may be helpful.

- Type 1, the chronic form of Gaucher's disease, is the most common. It results in an enlarged liver and spleen and bone abnormalities. Most commonly diagnosed during adulthood, type 1 Gaucher's disease may lead to severe liver disease, including increased risk of bleeding from the stomach and esophagus and liver cancer. Nervous system problems can also occur.
- Type 2, the infantile form, usually causes death in the first year of life. Affected infants have an enlarged spleen and severe nervous system problems.
- Type 3, the juvenile form, can begin at any time during childhood. Children with type 3 disease have an enlarged liver and spleen, bone abnormalities, and slowly progressive nervous system problems. Children who survive to adolescence may live for many years.

Many people with Gaucher's disease can be treated with enzyme replacement therapy, in which enzymes are given by vein, usually every 2 weeks. Enzyme replacement therapy is most effective for people who do not have nervous system complications.

TAY-SACHS DISEASE

Tay-Sachs disease is caused by a buildup of gangliosides in the tissues. This disease results in early death.

In Tay-Sachs disease, gangliosides, which are products of fat metabolism, accumulate in tissues. The disease is most common among families of Eastern European Jewish origin. At a very early age, children with this disease become progressively intellectually disabled and appear to have floppy muscle tone. Spasticity develops and is followed by paralysis, dementia, and blindness. These children usually die by age 3 or 4. The disease cannot be treated or cured.

Before conception, parents can find out whether they carry the gene that causes the disease. During pregnancy, Tay-Sachs disease can be identified in the fetus by chorionic villus sampling or amniocentesis.

NIEMANN-PICK DISEASE

Niemann-Pick disease is caused by a buildup of sphingomyelin or cholesterol in the tissues. This disease causes many neurologic problems.

In Niemann-Pick disease, the deficiency of a specific enzyme results in the accumulation of sphingomyelin (a product of fat metabolism) or cholesterol. Niemann-Pick disease has several forms, depending on the severity of the enzyme deficiency, which determines how much sphingomyelin or cholesterol accumulates. The most severe forms tend to occur in Jewish people. The milder forms occur in all ethnic groups.

In the most severe form (type A), children fail to grow normally and have several neurologic problems. These children usually die by age 3. Children with type B disease develop fatty growths in the skin, areas of dark pigmentation, and an enlarged liver, spleen, and lymph nodes. They may be intellectually disabled. Children with type C disease develop symptoms during childhood, with seizures and neurologic deterioration.

Some forms of Niemann-Pick disease can be diagnosed in the fetus by chorionic villus sampling or amniocentesis. After birth, the diagnosis can be made by a liver biopsy (removal of a tissue specimen for examination under a microscope). None of the types of Niemann-Pick disease can be cured, and children tend to die of infection or progressive dysfunction of the central nervous system. Currently, some therapies that may slow or halt the progression of symptoms in types B and C are being studied.

FABRY'S DISEASE

Fabry's disease is caused by a buildup of glycolipid in tissues. This disease causes skin growths, pain in the extremities, poor vision, recurrent episodes of fever, and kidney or heart failure.

In Fabry's disease, glycolipid, which is a product of fat metabolism, accumulates in tissues. Because the defective gene for this rare disorder is carried on the X chromosome, the full-blown disease occurs only in males (Genetics: X-Linked Inheritance).

The accumulation of glycolipid causes noncancerous (benign) skin growths (angiokeratomas) to form on the lower part of the trunk. The corneas become cloudy, resulting in poor vision. A burning pain may develop in the arms and legs, and children may have episodes of fever. Children with Fabry's disease eventually develop kidney failure and heart disease, although most often, they live into adulthood.

Kidney failure may lead to high blood pressure, which may result in stroke. Fabry's disease can be diagnosed in the fetus by chorionic villus sampling or amniocentesis. The disease cannot be cured or even treated directly, but researchers are investigating a treatment in which the deficient enzyme is replaced by transfusion. Treatment consists of taking analgesics to help relieve pain and fever or anticonvulsants. People with kidney failure may need a kidney transplant.

FATTY ACID OXIDATION DISORDERS

Fatty acid oxidation disorders are caused by a lack or deficiency of the enzymes needed to break down fats, resulting in delayed mental and physical development.

Several enzymes help break down fats so that they may be turned into energy. An inherited defect or deficiency of one of these enzymes leaves the body short of energy and allows breakdown products, such as acyl-CoA, to accumulate. The enzyme most commonly deficient is medium chain acyl-CoA dehydrogenase (MCAD). Other enzyme deficiencies include short chain acyl-CoA-dehydrogenase deficiency (SCAD), long chain-3-hydroxyacyl-CoA-deficiency (LCHAD), and trifunctional protein deficiency (TFP). MCAD Deficiency: This disorder is one of the most common inherited disorders of metabolism, particularly among people of Northern European descent.

Symptoms usually develop between birth and age 3. Children are most likely to develop symptoms if they go without food for a period of time (which depletes other sources of energy) or have an increased need for calories because of exercise or illness. The level of sugar in the blood drops significantly, causing confusion or coma. Children become weak and may have vomiting or seizures. Over the long term, children have delayed mental and physical development, an enlarged liver, heart muscle weakness, and an irregular heartbeat. Sudden death may occur.

Since 2007, nearly every state in the United States has screened all newborns for MCAD with a blood test. Immediate treatment is with glucose given by vein. For long-term treatment, children must eat often, never skip meals, and consume a diet high in carbohydrates and low in fats. Supplements of the amino acid carnitine may be helpful. The long-term outcome is generally good

Chapter 14

Aminoacids and Proteins

PROTEINS

Proteins are any of a group of complex organic compounds containing carbon, hydrogen, nitrogen and sulfur. Proteins, the principle constituents of the protoplasm of all cells, are of high molecular weight and consist of alpha-amino acids joined by peptide linkages. Different amino acids are commonly found in proteins, each protein having a unique, genetically defined amino-acid sequence, which determines its specific shape and function. They serve as enzymes, structural elements, hormones, immunoglobulins, etc. and are involved in oxygen transport, muscle contraction, electron transport, and other activities.

The importance of adequate protein intake to proper immune function has been extensively studied. The most severe effects of"Protein-Calorie Malnutrition" (PCM) are on cell-mediated immunity, although all facets of immune function are ultimately affected. PCM is not, however, a single nutrient deficiency. It is normally associated with multiple nutrient deficiencies, and some immune dysfunctions attributed to PCM are most likely due to these other factors. Partial deficiencies of dietary vitamins produce a comparatively greater depression on immune functions than do partial protein deficiencies. Non-the less, adequate protein is essential for optimal immune function.

High protein diets are not recommended for individuals with kidney or liver disease. There are no other known side effects.

A macromolecule made from amino acids; some consist of one, long polypeptide chain while others are made from several intertwined polypeptides (proteios = first place)

Proteins are about 50% of the dry weight of most cells, and are the most structurally complex macromolecules known. Each type of protein has its own unique structure and function.

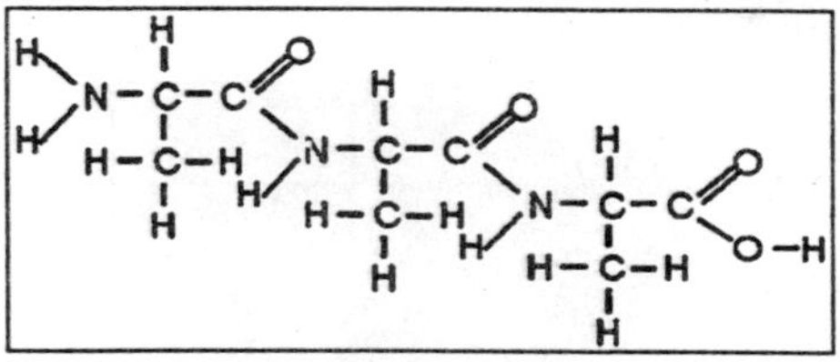

POLYMER

A large molecule made of many, smaller, repeating, identical or similar subunits (poly = many; mer = a part) Polymers are any kind of large molecules made of repeating identical or similar subunits called monomers. The starch and cellulose we previously discussed are polymers of glucose, which in that case, is the monomer. Proteins are polymers of about 20 amino acids (the monomer).

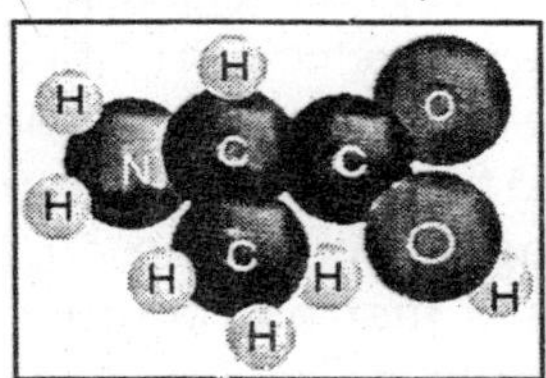

Amino acids are built from a central carbon bonded to four different groups. (run your mouse over the amino acid to see the names of the parts).

- Hydrogen (–H),
- Amino group (–NH_2),
- Carboxyl group (–COOH),

and some side chain symbolized by"R".

In the approximately-20 amino acids found in our bodies, what varies is the side chain. Some side chains are hydrophilic while others are hydrophobic.

Since these side chains stick out from the backbone of the

molecule, they help determine the properties of the protein made from them.

Most naturally-occurring amino acids are the l- form, whereas synthetically-produced amino acids give a 50:50 mixture. Notice that these molecules are mirror images of each other, thus there is no way you can rotate one molecule to make it look like the other.

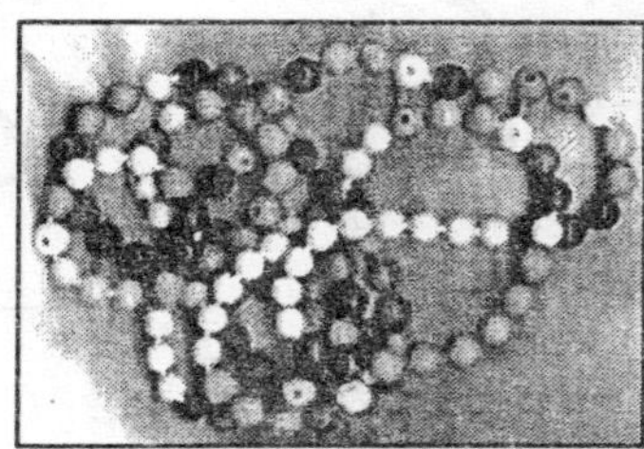

To form protein, the amino acids are linked by dehydration synthesis to form peptide bonds. The chain of amino acids is also known as a polypeptide. Some proteins contain only one polypeptide chain while others, such as hemoglobin, contain several polypeptide chains all twisted together. The sequence of amino acids in each polypeptide or protein is unique to that protein, so each protein has its own, unique 3-D shape or native conformation. If even one amino acid in the sequence is changed, that can potentially change the protein's ability to function. For example, sickle cell anemia is caused by a change in only one nucleotide in the DNA sequence that causes just one amino acid in one of the hemoglobin polypeptide molecules to be different. Because of this, the whole red blood cell ends up being deformed and unable to carry oxygen properly.

```
         H
H\       |    //O
  N - C - C         H
H/       |    \     |    //O
     H - C - H  N - C - C          H
         |     /    |    \         |    //O
         H    H H - C - H  N - C - C
                    |     /    |    \
                    H    H H - C - H  O - H
                               |
                               H
```

As a polypeptide chain forms, it naturally twists and bends into its native conformation. One of the things that helps determine the native conformation of a protein is the side chains of all the amino acids involved. Remember some amino acid side chains are hydrophobic while others are hydrophilic. In this case, the"likes" attract: all the hydrophobic side chains (here represented by yellow beads) try to"get together" in the Centre of the molecule, away from the watery environment, while the hydrophilic side chains are attracted to the outside of the molecule, near the watery environment. Additionally, some of the hydrophilic side chains have groups of atoms attached that make them acidic, while others have groups attached that make them basic. Side chains with acidic ends are attracted to side chains with basic ends, and can form ionic bonds. Thus, the side chains interacting with each other help to hold the protein in its native conformation.

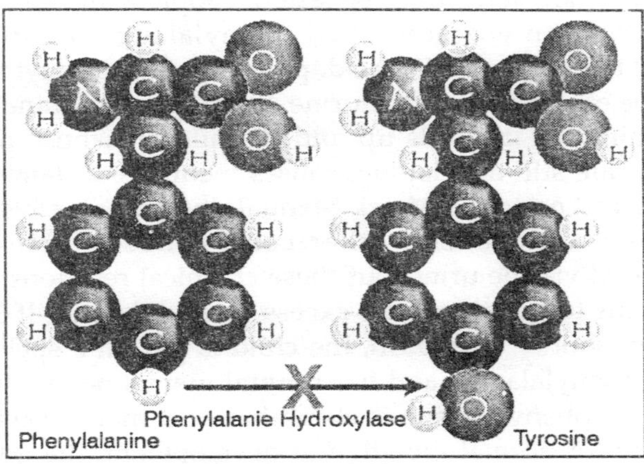

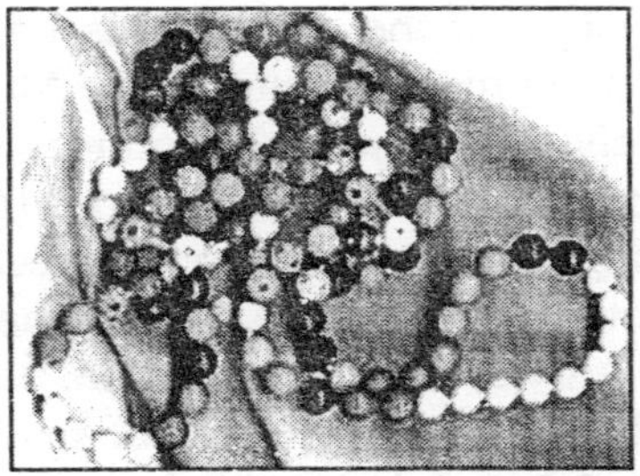

Denaturation is when a protein loses its native conformation. For example, egg white (also called albumen)

contains a protein called albumin which is water-soluble. However, if heated, albumin becomes denatured and loses its ability to be water-soluble. There are several possible things that can denature proteins including changing the temperature (adding heat), changing the pH or salt concentration of the solution, or putting the protein into a hydrophobic solvent. In a hydrophobic solvent, the amino acids with hydrophobic side chains (the yellow beads) would all try to go to the outside of the molecule, and all those with hydrophilic side chains would cluster in the Centre of the molecule. If a protein remains water-soluble when denatured (unlike albumin), it can return to its native conformation if/when placed back into a"normal" environment.

PKU or phenylketonuria is a genetic disorder for which, by law, all newborns must be tested. People with this disorder are lacking an enzyme called phenylalanine hydroxylase needed to turn the amino acid, phenylalanine into tyrosine so it can be eliminated via the urine. Since phenylalanine cannot be eliminated, it builds up and in children whose nervous systems are still developing, causes mental retardation.

According to the Merck Manual, the body is slowly able to turn it into several other compounds which can be eliminated via the urine, but these chemical reactions are too inefficient to get rid of the excess phenylalanine. If PKU is detected shortly after birth, the child is put on a special diet low in phenylalanine and has normal mental development.

Since phenylalanine is an amino acid that is needed by the body for normal growth and development, the goal is to limit intake to just what the child needs without any excess that could build up in his/her system. The Merck Manual says that, unfortunately, most natural protein sources, including milk, are too high in phenylalanine for children with PKU, so these babies/children must be put on a special formula from which most, if not all, of the phenylalanine has been removed.

As the child grows, some low-protein, natural foods such as fruits, vegetables, and some kinds of grain are OK to eat. The needed phenylalanine is supplied via carefully-measured amounts of natural protein plus the residual in the formula.

One area of"debate" is how long the special diet must be continued. Some people think the diet may be discontinued after the brain finishes developing at about age 5. Others have noted some Behaviour problems and learning difficulties that appear to be related to going off the diet, and recommend that people with PKU stay on the diet for life.

Staying on the special diet is especially important if a woman who has PKU is even thinking about trying to become pregnant. Because any excess phenylalanine in her system could cross the placenta and could cause mental retardation in the baby, especially in the first few weeks while the nervous system is forming, she has to make sure that she adheres to the special diet from before conception. Note that this is not a major concern for a normal woman carrying a baby with PKU (she wouldn't know that until after the baby was born and tested) because the woman's body can digest phenylalanine, including any excess from the baby's body that is sent back, via the placenta, to her body.

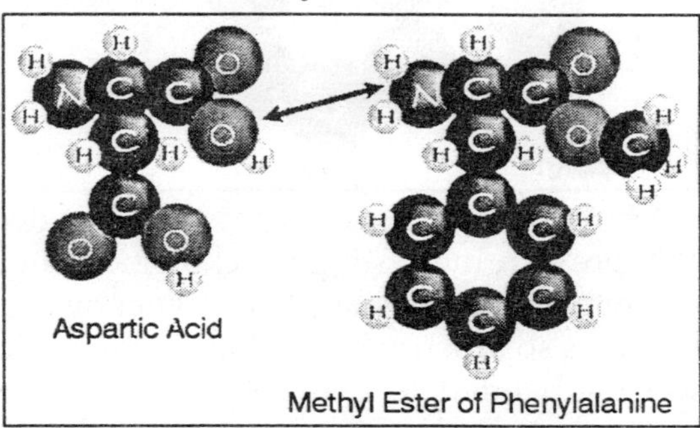

Nutrasweet has the chemical name aspartame and is made of two amino acids: l-aspartic acid and the methyl ester of phenylalanine, thus people with PKU can't have it. Look at a can of diet soft drink that is sweetened with aspartame. Somewhere on the can, there's always a cautionary label warning people with PKU not to drink it.

But what about the rest of us? That methyl group on the phenylalanine can break off pretty easily, and at temperatures as"low" as 86° F (= 30° C - lower than body temperature or a summer day), aspartame starts to disintegrate, releasing methanol (which in larger quantity causes blindness, etc. - it's toxic). As the liver tries to deal with the methanol using normal metabolic pathways, it ends up turning the methanol into formaldehyde, which is a known carcinogen. Interestingly, because of this and several other concerns, originally there was controversy over FDA approval of aspartame. Even still, its safety is controversial. More recently, some people have suggested a possible link between aspartame consumption and an increase in brain tumors in this country.

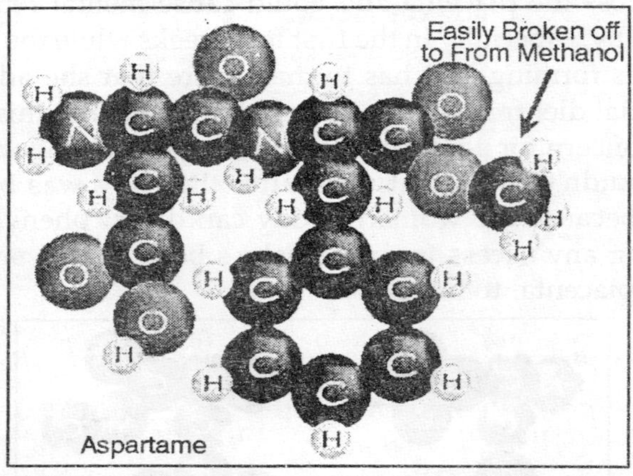

Equal® is not the same as Nutrasweet, but does contain some of it. There's an interesting psychological thing here: in our culture, people think they need sweetener by the teaspoon, but Nutrasweet is so much sweeter than sugar that only a tiny pinch would be needed to gain the equivalent sweetness. Artificial sweetner manufacturers know that if they don't make a product such that one teaspoon is as sweet as one teaspoon of the sugar people are used to using, people will use one teaspoon anyway (a tiny pinch just doesn't look like enough), then say it was too sweet and stop buying it. Thus, artificial sweetners are designed to take up about a teaspoon of space, and Equal® is mostly glucose (which diabetics and hypoglycemics shouldn't have) with a little Nutrasweet added, so in theory, not quite as much of the sugar is needed (also remember that glucose isn't as sweet as sucrose).

To make the sweetness of one teaspoon of sweetner

powder equivalent to that of one teaspoon of sucrose, various fillers (things like cellulose) are added. These products also contain a number of other chemicals, including silicon dioxide (SiO2)"to prevent caking." This is the chemical name for sand or glass (although, from what I've heard, a bit of silicon is needed for healthy fingernails)! Pre-mixed, pre-packaged items, such as artificially-sweetened soft drinks, can contain all aspartame as their sweetener because the manufacturers of these products do not have to be concerned with the"one teaspoon" psychological factor.

Remember, because aspartame degrades into methanol at warm temperatures, it should not be used in anything hot or anything that will be cooked (coffee sweetener, cake mixes, cocoa, etc.), and bottles of artificially-sweetened soft drinks that have been sitting out in the hot sun at the local gas station are suspect.

STRUCTURE OF PROTEINS

This explains how amino acids combine to make proteins and what is meant by the primary, secondary and tertiary structures of proteins. Quaternary structure isn't covered. It only applies to proteins consisting of more than one polypeptide chain. There is a mention of quaternary structure on the IB chemistry syllabus, but on no other UK-based syllabus at this level.

Quaternary structure can be very complicated, and I don't know exactly what depth the IB syllabus wants for this (which is why I haven't included it). I suspect what is wanted is fairly trivial. IB students should ask the advice of their teacher or lecturer.

PRIMARY STRUCTURE OF PROTEINS

Drawing the Amino Acids

In chemistry, if you were to draw the structure of a general 2-amino acid, you would probably draw it like this:

$$\begin{array}{c} \quad NH_2 \\ \quad | \\ R-CH-COOH \end{array}$$

However, for drawing the structures of proteins, we usually twist it so that the"R" group sticks out at the side. It is much easier to see what is happening if you do that.

$$\begin{array}{c} R \\ | \\ NH2-CH-COOH \end{array}$$

That means that the two simplest amino acids, glycine and alanine, would be shown as:

$$\begin{array}{c} H \\ | \\ NH_2-CH-COOH \end{array}$$

Glycine

$$\begin{array}{c} CH_3 \\ | \\ NH_2-CH-COOH \end{array}$$

alanine

PEPTIDES AND POLYPEPTIDES

Glycine and alanine can combine together with the elimination of a molecule of water to produce a dipeptide. It is possible for this to happen in one of two different ways - so you might get two different dipeptides.

Either:

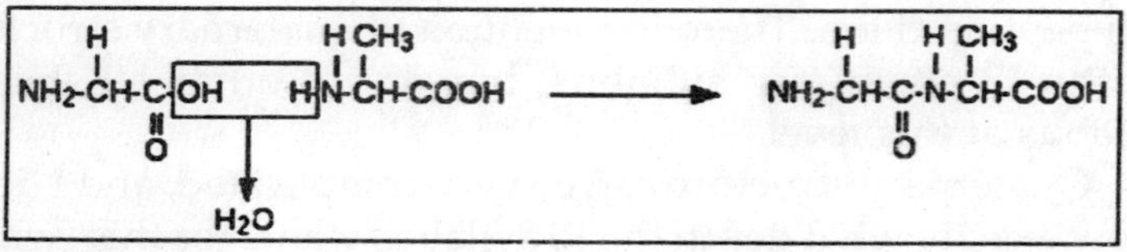

$$NH_2\text{-}CH(H)\text{-}C(=O)\text{-}OH + H\text{-}N(H)\text{-}CH(CH_3)\text{-}COOH \longrightarrow NH_2\text{-}CH(H)\text{-}C(=O)\text{-}N(H)\text{-}CH(CH_3)\text{-}COOH + H_2O$$

Or:

$$NH_2\text{-}CH(CH_3)\text{-}C(=O)\text{-}OH + H\text{-}N(H)\text{-}CH(H)\text{-}COOH \longrightarrow NH_2\text{-}CH(CH_3)\text{-}C(=O)\text{-}N(H)\text{-}CH(H)\text{-}COOH + H_2O$$

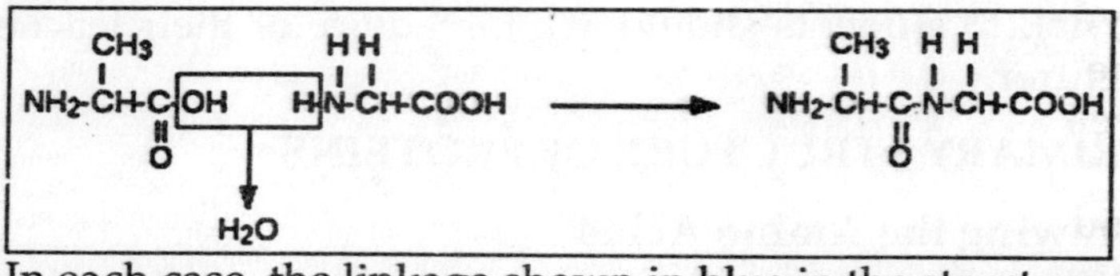

In each case, the linkage shown in blue in the structure of the dipeptide is known as a peptide link. In chemistry, this would also be known as an amide link, but since we are now in the realms of biochemistry and biology, we'll use their terms.

If you joined three amino acids together, you would get a

tripeptide. If you joined lots and lots together (as in a protein chain), you get a polypeptide.

A protein chain will have somewhere in the range of 50 to 2000amino acid residues. You have to use this term because strictly speaking a peptide chain isn't made up of amino acids. When the amino acids combine together, a water molecule is lost. The peptide chain is made up from what is left after the water is lost - in other words, is made up of amino acid residues. By convention, when you are drawing peptide chains, the -NH_2group which hasn't been converted into a peptide link is written at the left-hand end. The unchanged –COOH group is written at the right-hand end. The end of the peptide chain with the –NH_2 group is known as theN-terminal, and the end with the –COOH group is the C–

Terminal

A protein chain (with the N-terminal on the left) will therefore look like this:

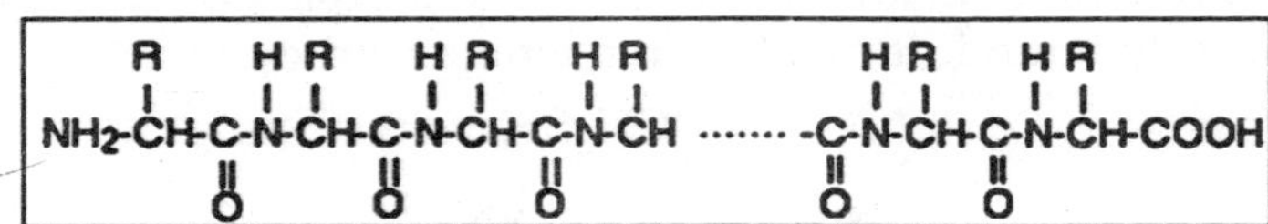

The"R" groups come from the 20 amino acids which occur in proteins. The peptide chain is known as the backbone, and the"R" groups are known as side chains.

Note: In the case where the"R" group comes from the amino acid proline, the pattern is broken. In this case, the hydrogen on the nitrogen nearest the"R" group is missing, and the"R" group loops around and is attached to that nitrogen as well as to the carbon atom in the chain.

I mention this for the sake of completeness - not because you would be expected to know about it in chemistry at this introductory level.

At its simplest, the term is used to describe the order of the amino acids joined together to make the protein. In other words, if you replaced the"R" groups in the last diagram by real groups you would have the primary structure of a particular protein.

This primary structure is usually shown using abbreviations for the amino acid residues. These abbreviations commonly consist of three letters or one letter.

Using three letter abbreviations, a bit of a protein chain might be represented by, for example:

Try.Gly.Lys.pro.Val.Gly. Lys.Lys.Arg.Arg.Pro.Val.Lys.
Tyr.Pro.Ala.Gly.Glu...

If you look carefully, you will spot the abbreviations for glycine (Gly) and alanine (Ala) amongst the others.

If you followed the protein chain all the way to its left-hand end, you would find an amino acid residue with an unattached $-NH_2$ group. The N-terminal is always written on the left of a diagram for a protein's primary structure - whether you draw it in full or use these abbreviations.

The wider definition of primary structure includes all the features of a protein which are a result of covalent bonds. Obviously, all the peptide links are made of covalent bonds, so that isn't a problem.

But there is an additional feature in proteins which is also covalently bound. It involves the amino acid cysteine.

$$\begin{array}{c} CH_2SH \\ | \\ NH_2-CH-COOH \end{array}$$

Cysteine

If two cysteine side chains end up next to each other because of folding in the peptide chain, they can react to form a sulphur bridge. This is another covalent link and so some people count it as a part of the primary structure of the protein.

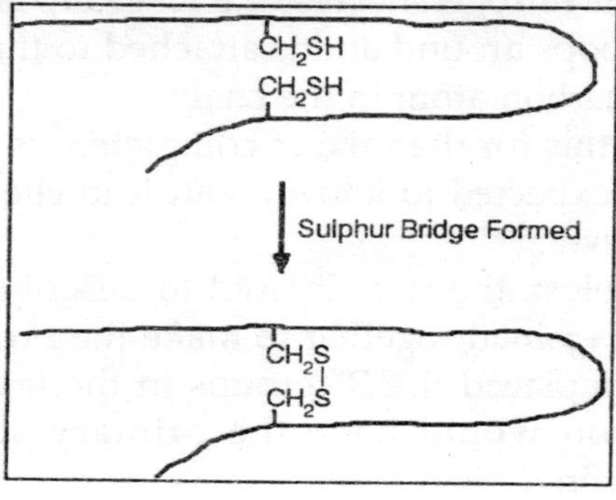

Because of the way sulphur bridges affect the way the protein folds, other people count this as a part of the tertiary structure (see below). This is obviously a potential source of confusion!

Important: You need to know where your particular examiners are going to include sulphur bridges - as a part of the primary structure or as a part of the tertiary structure. You need to check your current syllabus and past papers. If you are studying a UK-based syllabus and haven't got these, follow this link to find out how to get hold of them.

Secondary Structure of Proteins

Within the long protein chains there are regions in which the chains are organised into regular structures known as alpha-helices (alpha-helixes) and beta-pleated sheets. These are the secondary structures in proteins.

These secondary structures are held together by hydrogen bonds. These form as shown in the diagram between one of the lone pairs on an oxygen atom and the hydrogen attached to a nitrogen atom:

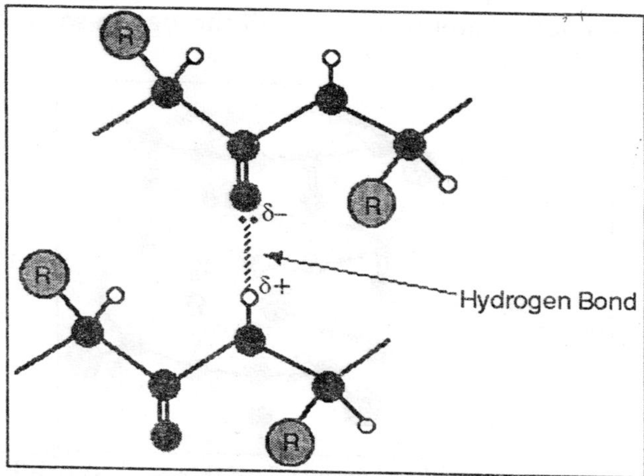

Important: If you aren't happy about hydrogen bonding and are unsure about what this diagram means, follow this link before you go on.

What follows is difficult enough to visualise anyway

without having to worry about what hydrogen bonds are as well!

You must also find out exactly how much detail you need to know about this next bit. It may well be that all you need is to have heard of an alpha-helix and know that it is held together by hydrogen bonds between the C=O and N-H groups. Once again, you need to check your syllabus and past papers - particularly mark schemes for the past papers.

The Alpha-helix

In an alpha-helix, the protein chain is coiled like a loosely-coiled spring. The"alpha" means that if you look down the length of the spring, the coiling is happening in a clockwise direction as it goes away from you.

Note: If your visual imagination is as hopeless as mine, the only way to really understand this is to get a bit of wire and coil it into a spring shape. The lead on your computer mouse is fine for doing this!

The next diagram shows how the alpha-helix is held together by hydrogen bonds. This is a very simplified diagram, missing out lots of atoms. We'll talk it through in some detail after you have had a look at it.

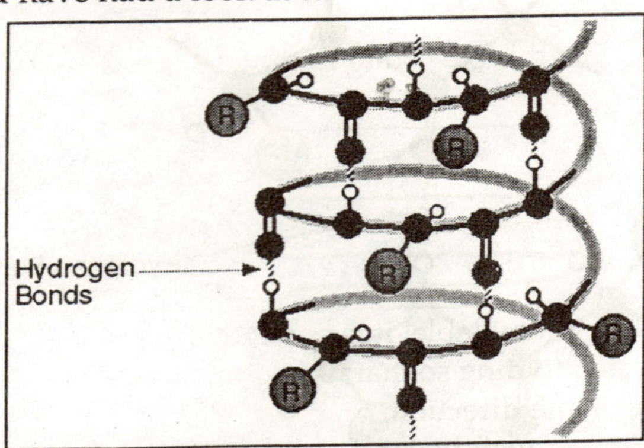

What's wrong with the diagram? Two things: First of all, only the atoms on the parts of the coils facing you are shown. If you try to show all the atoms, the whole thing gets so complicated that it is virtually impossible to understand what is going on.

So, what do you need to notice? Notice that all the"R" groups are sticking out sideways from the main helix. Notice the regular arrangement of the hydrogen bonds. All the N-H groups are pointing upwards, and all the C=O groups pointing downwards.

Each of them is involved in a hydrogen bond. And finally, although you can't see it from this incomplete diagram, each complete turn of the spiral has 3.6 (approximately) amino acid residues in it.

If you had a whole number of amino acid residues per turn, each group would have an identical group underneath it on the turn below. Hydrogen bonding can't happen under those circumstances.

Each turn has 3 complete amino acid residues and two atoms from the next one. That means that each turn is offset from the ones above and below, such that the N-H and C=O groups are brought into line with each other.

Beta-pleated Sheets

In a beta-pleated sheet, the chains are folded so that they lie alongside each other. The next diagram shows what is known as an"anti-parallel" sheet.

All that means is that next-door chains are heading in opposite directions.

Given the way this particular folding happens, that would seem to be inevitable.

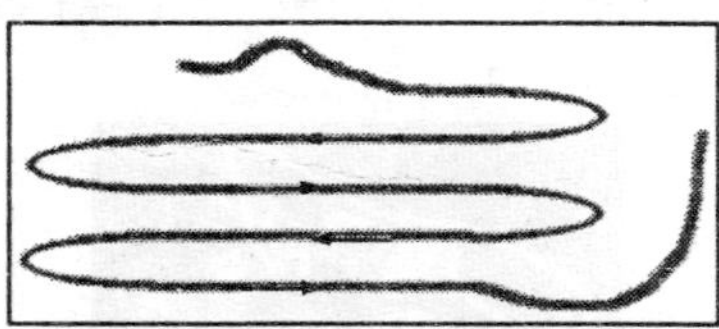

It isn't, in fact, inevitable! It is possible to have some much more complicated folding so that next-door chains are actually heading in the same direction.

We are getting well beyond the demands of UK A level chemistry (and its equivalents) now. The folded chains are again held together by hydrogen bonds involving exactly the same groups as in the alpha-helix.

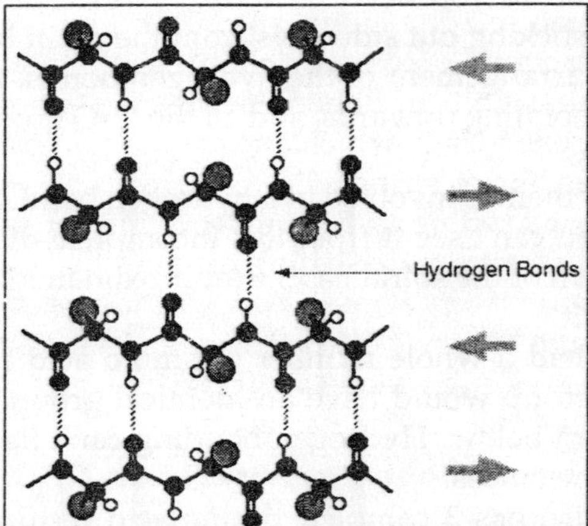

Note: Note that there is no reason why these sheets have to be made from four bits of folded chain alongside each other as shown in this diagram. That was an arbitrary choice which produced a diagram which fitted nicely on the screen!

TERTIARY STRUCTURE OF PROTEINS

Definition

The tertiary structure of a protein is a description of the way the whole chain (including the secondary structures) folds itself into its final 3-dimensional shape. This is often simplified into models like the following one for the enzyme dihydrofolate reductase. Enzymes are, of course, based on proteins.

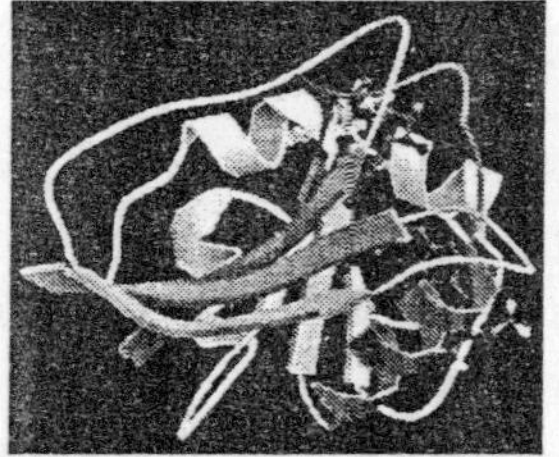

Note: This diagram was obtained from the RCSB Protein Data Bank. If you want to find more information about dihydrofolate reductase, their reference number for it is 7DFR. There is nothing particularly special about this enzyme in terms of structure. The model shows the alpha-helices in the secondary structure as coils of"ribbon". The beta-pleated sheets are shown as flat bits of ribbon ending in an arrow head. The bits of the protein chain which are just random coils and loops are shown as bits of"string".

The colour coding in the model helps you to track your way around the structure - going through the spectrum from dark blue to end up at red. You will also notice that this particular model has two other molecules locked into it (shown as ordinary molecular models). These are the two molecules whose reaction this enzyme catalyses.

Holding Protein into its Tertiary Structure

The tertiary structure of a protein is held together by interactions between the the side chains - the"R" groups. There are several ways this can happen.

Ionic Interactions

Some amino acids (such as aspartic acid and glutamic acid) contain an extra -COOH group. Some amino acids (such as lysine) contain an extra $-NH_2$ group.

You can get a transfer of a hydrogen ion from the -COOH to the -NH2 group to form zwitterions just as in simple amino acids. You could obviously get an ionic bond between the negative and the positive group if the chains folded in such a way that they were close to each other.

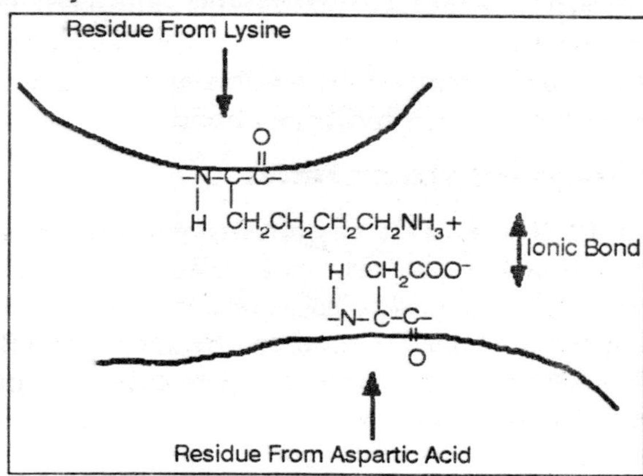

Hydrogen Bonds

Notice that we are now talking about hydrogen bonds between side groups - not between groups actually in the backbone of the chain.

Lots of amino acids contain groups in the side chains which have a hydrogen atom attached to either an oxygen or a nitrogen atom. This is a classic situation where hydrogen bonding can occur. For example, the amino acid serine contains an -OH group in the side chain. You could have a hydrogen bond set up between two serine residues in different parts of a folded chain.

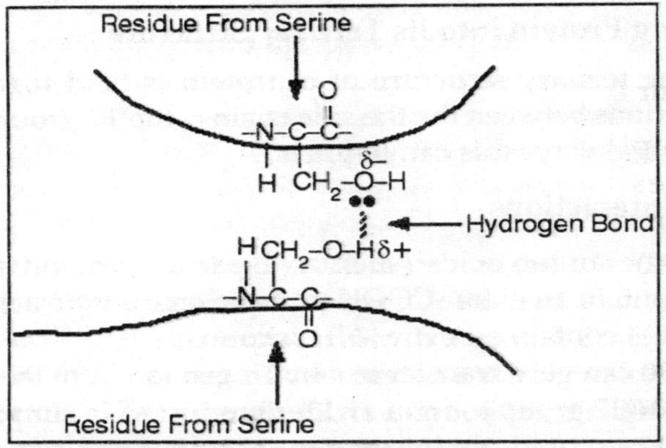

You could easily imagine similar hydrogen bonding involving -OH groups, or -COOH groups, or -CONH2 groups, or -NH2 groups in various combinations - although you would have to be careful to remember that a -COOH group and an -NH2 group would form a zwitterion and produce stronger ionic bonding instead of hydrogen bonds.

Van der Waals Dispersion Forces

Several amino acids have quite large hydrocarbon groups in their side chains. A few examples are shown below. Temporary fluctuating dipoles in one of these groups could induce opposite dipoles in another group on a nearby folded chain. The dispersion forces set up would be enough to hold the folded structure together.

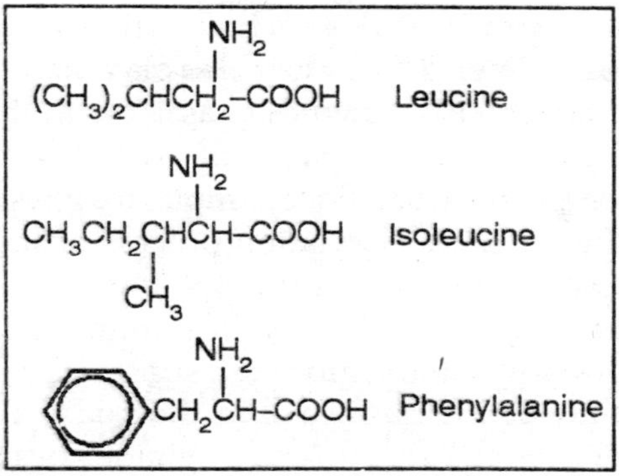

NH_2
$(CH_3)_2CHCH_2\text{-}COOH$ Leucine

NH_2
$CH_3CH_2CHCH\text{-}COOH$ Isoleucine
CH_3

NH_2
$C_6H_5CH_2CH\text{-}COOH$ Phenylalanine

Sulphur Bridges

Sulphur bridges which form between two cysteine residues have already been discussed under primary structures. Wherever you choose to place them doesn't affect how they are formed.

AMINO ACIDS

One of a class of organic compounds containing the amino (NH_2) and the carboxyl (COOH) group, occurring naturally in plant and animal tissues and forming the chief constituents of protein; many of them are necessary for human and animal growth and nutrition and hence are called Essential Amino Acids. Amino Acids are the building blocks of protein.

PRIMARY FUNCTIONS OF AMINO ACIDS

Depending upon the source of dated reference material studied, some reports state"20" Amino Acids have been identified. More modern reference materials state"22" Amino Acids have been identified. In my personal research, identifying all possibilities available to me at this time, I have listed"28" Amino Acids/ or Amino Acid Derivatives. Of my 28 listings, 22 have been identified as"Commonly Occurring" in Proteins.

In living things, Amino Acids serve as the primary unit of synthesis both of Tissue Proteins and other Non-Protein Nitrogenous Compounds.

Amino acids may undergo complex inner conversions in the body into macromolecules such as carbohydrates, nucleic acids, lipids, etc. Mineral co-factors also play an important role in that they interact with enzymes to assist with their catalytic activities. They are also capable of complexing with the individual free amino acids. For example, magnesium (MG2+) is important in the phosphorylation process or the transfer of high-energy phosphate groups into the body.

Upon entering the body any free amino acids undergo complex inner-conversions into those substances necessary to maintain the metabolic pathways in the body. For example, Phenylalanine goes to Tyrosine and Arginine goes to Citrulline and Ornithine. (The Arginine pathway is a very important cycle.) Amino acid inner-conversions are not limited to other amino acids. Many free form amino acids play important roles as precursors for metabolic intermediaries.

An example of this, Tyrosine's and therefore Phenylalanine's ability to give rise to two hormones Thyroxsine and Epinephrine. The Sulfur containing amino acids often precurse Taurine, which is an important bio-acid component. As W.C. Rose observed in 1938,"Perhaps undo emphasis was placed upon the quantity of protein ingestion and that scant consideration was given to possible difference in nutrient quality."

Upon ingestion of foodstuff all nutrients and amino acids do not immediately diffuse into the surrounding tissue of the blood stream, but must first undergo a series of steps of biochemical reactions in the digestive tract. These reactions reduce the protein into its' individual amino acids. Protein cleaving enzymes are activated in the digestive process, but in order for this cleavage to occur energy must be supplied on the part of the individuals metabolism to break the peptide bonds. By starting with a free amino acid mixture in which all amino acids are in their L conformation little digestion is necessary. The result is an energy saving and rapid absorption into the blood stream and surrounding tissues."Amino acid supplementation of foods and feeds bring about many beneficial results. Of practical importance is the attainment of a better balance of protein and therefore a better balanced diet, an extension or savings of the available protein supply, and an improvement in the efficiency of the protein and food utilization.

FUNCTIONAL IDENTITY OF AMINO ACIDS

Of all the fundamental amino acids, 10 (ten) are classified as"ESSENTIAL" and are NOT biosynthesized by the body. All commonly occurring amino acids can be subdivided into 5 fundamental groups each of which has functional significance.

- *Group* #1: This group includes all the amino acids possessing aliphatic properties (those whose carbon atoms are joined in open chains), Glycine, Alanine, Leucine, w, Serine, Threonine, and Isoleucine.
- *Group* #2: These are the Sulfur containing amino acids, Methionine, Cystine, Cysteine.
- *Group* #3: These amino acids are the acidic amino acids, (those that can donate a hydrogen ion), Aspartic Acid, Glutamic Acid, and their associated amids.
- *Group* #4: These are the diamino acids (those with additional NH_2 groups), Arginine and Lysine.
- *Group* #5: These are the cyclic and aromatic amino acids, Phenylalanine, Tyrosine, Tryptophan, Histidine, Proline, and Hydroxyproline that is synthesized from Prolamine.

ESSENTIAL AMINO ACIDS

Table. Estimated Minimum Requirements of Adults for Essential Amino Acids

Amino Acid	Women/gms/day	Men/gms/day
Cystine	0.055	1.010
Isoleucine	0.450	0.700
Leucine	0.620	1.100
Lysine	0.500	0.800
Methionine	questionable	1.100
Phenylalanine	0.220	1.010

Threonine	0.310	0.500
Tryptophan	0.160	0.250
Tyrosine	0.900	questionable
Valine	0.650	0.800

Cystine

An essential, Sulfur-containing amino acid, produced by digestion or acid hydrolysis of proteins, sometimes found in the urine and kidneys, and readily reduced to two molecules of Cysteine.

- Reported helpful in dermatological conditions.
- Promotes faster recovery of tissue after surgery.
- Part of the Insulin molecule.
- Found high in hair (sulfur bonds).

Isoleucine

An essential amino acid produced by hydrolysis of fibrin and other proteins; ESSESNTIAL for optimal INFANT growth and for nitrogen equilibrium in adults.

- Needed along with other branch chain amino acids for all rebuilding of muscle tissue.
- Role in the release of energy during muscular work. (Metabolizes along the same pathway as fat.)

Leucine

An essential amino acid, ESSENTIAL for optimal growth in INFANTS, and for nitrogen equilibrium in adults.

- Metabolized along the same pathways as fat.
- Precursors of cholesterol.
- Involved in the role of energy release during any work of the muscles.

Lysine

An essential, naturally occurring amino acid, ESSESNTIAL for optimal growth in human INFANTS, and for maintenance of nitrogen equilibrium in adults.

- Reported to inhibit growth and replication of herpes simplex and Epstein Barr viruses (EBV).

- Promotes bone growth in infants.
- Stimulates secretion of gastric juices.
- Found in abundance in muscle tissue, connective tissue and collagen.
- Low in vegetarian diets.

Methionine

An essential, naturally occurring amino acid, which is an essential component of the diet, furnishing both methyl groups and sulfur necessary for normal metabolism.

- Prevents deposits and cohesion of fats in the liver.
- Gives rise to Taurine (an important inhibitory neuro modulator in the brain).
- Involved in synthesis of Choline (Must be given with Vitamin B6 (paradoxial-5-phosphate) to inhibit synthesis of homocysteine which promotes plaque deposition in the arteries.)

Phenylalanine

An essential, naturally occurring amino acid, ESSESNTIAL for optimal growth in INFANTS, and for nitrogen equilibrium in human adults.

- May be useful in appetite control by stimulating CCK (Chaleceptokinin Enzyme) secretion.
- Shown to be useful in management of certain types of depression.
- Increases blood pressure in hypotension.
- Gives rise to Tyrosine.
- It is one of the amino acids in the dypeptide sweetener Aspartain. (Aspartic acid and Phenolalanine).

Threonine

An essential, naturally occurring amino acid, essential for human metabolism.

- Rises to three times it's normal value at pregnancy.
- Acts as a lipotropic factor.
- Recently found to increase brain Glycine content,

greatly reducing ALS symptoms. (Amyotrophic Lateral Sclerosis or"Lue Gehrig's Disease.")

Tryptophan

An essential, naturally occurring amino acid, existing in proteins and essential for human metabolism.

- Reported useful in the management of depression and schizophrenia.
- Produces Serotonin, which induces sleep. Has a Serotonengenic effect.
- Precursor of the vitamin Niacin.
- Vasoconstrictor, which appears to aid in blood clotting mechanism, aids in elevating the threshold of pain.

Tyrosine

An essential, naturally occurring amino acid present in most proteins. It is a product of Phenylalanine metabolism and a precursor of thyroid hormones, catecholamines, and melanin.

- Along with Phenylalanine, it is a useful anti depressant due to increased production of Tacolamine. (Recommended to be taken with Vitamin B6 (paradoxial-5-phosphate).)
- Reported to stabilize blood pressure by lowering energy in some cases and elevating it in others
- Involved in tissue pigmentation
- Is important in the formation of thyroid hormone

Valine

An essential, naturally occurring amino acid, essential for human metabolism.

- Needed for all muscle building.
- Required in the precursors of cholesterol.

Histidine

An amino acid obtainable from many proteins by the action of sulfuric acid and water; it is ESSENTIAL for optimal growth in INFANTS. Its decarboxylation results in formation of histamine.

- Release of histamines from body stores are required for sexual arousal.
- Reported useful in alleviating pain associated with rheumatoid arthritis. (Should be taken with vitamin C; Should be used with caution with Manic Depressives with elevated histamines; Should be used with caution in woman with severe depression or suicidal tendencies due to Premenstrual Syndrome; Is considered a mild neurotransmitter).

TYPES OF AMINO ACIDS

Aminoacetic Acid

A nonessential amino acid, derived from many proteins, used as a gastric antacid and dietary supplement, and in the treatment of various myopathies and peripheral vascular insufficiencies.

A solution is used as an irrigation fluid. Its hydrochloride salt is use as a source of hydrochloric acid in the treatment of achlorhydria.

Alanine

A natural amino acid occurring in two forms, alpha-Alanine, and beta-Alanine.

- Hypoglycemia (Prolongs the stabilization of blood glucose over long periods of time.)
- Alanine may be used for the production of glucose in gluconeogenesis.

Arginine

An amino acid occurring in proteins; it is involved in the urea cycle, which converts ammonia to urea.

- Induces growth hormone release from the pituitary gland.
- Major component of seminal fluid.
- Helpful in burn treatment, elevated ammonia levels, and cirrhosis of the liver.
- Stimulates immune response by enhancing the production of T-Cells.

- Protective effect of toxicity of hydrocarbons and intravenous diuretics.

(Should be used with caution in schizophrenic cases.)

Intake should be kept low in persons with Herpes simplex and Epstein Barr Virus (EBV. Mechanism of interaction is believed because of possible replications of the Herpes and EB viruses.)

Aspartic Acid

A nonessential, natural dibasic amino acid, involved in transamination reactions, the ornithine cycle, and the formation of carnosine, anserine, purines, and pyrimidines.

- Has a protective function over the liver.
- Helps in detoxification of ammonia.
- Promotes mineral uptake in the intestinal tract.
- It is part of the sweetener Aspartain that is a dypeptide of the amino acids; Aspartic Acid and Phenylalanine.

Carnitine

A betaine derivative involved in the transport of fatty acids into mitochondria, where they are metabolized.

- Suggested as useful in mobilization of surface fats such as cellulite.
- Helpful in treatment of fatigue and when muscle weakness is present. Useful in the oxidation of long chain fatty acids. A major source of energy for tissues.
- Reported as useful in myocardial ischemia (low energy metabolism by the heart).
- Useful in clearing triglycerides from the blood.
- Useful in metabolic abnormalities.

Citrulline

An alpha-amino acid involved in urea production; formed from ornithine and is itself converted into Arginine in the urea cycle.

- Produces the amino acids, Arginine and Ornithine.
- Detoxifies nitrogen containing waste products such

as ammonia (a cellular toxin) and is therefore part of ammonia detoxification (urea cycle).
- Stimulates growth hormone production.

Cysteine

A sulfur-containing amino acid produced by enzymatic or acid hydrolysis of proteins, readily oxidized to Cystine, sometimes found in urine.

- Assists tissues damaged by alcohol abuse, cigarette smoke and air pollution through the detoxification of Acetalaldahide.
- Helps maintain skin flexibility and texture by slowing abnormal cross linking of collagen (connective tissue protein which holds the skin together).
- Promotes red and white blood cell reproduction and tissue restoration in lung affecting diseases.
- Useful in iron deficient anemia and promotion of iron absorption.
- Helpful in prevention of peroxidized fats and free radicals.
- Converts to Cystinine in the absence of vitamin C.
- One of the amino acids found in tripeptide Glutathione and GTF.
- Hair contains between 10 to 16% Cysteine.
- Heavy metal chelator. (Should be used with caution in diabetics due to possible three dimensional structure changes in the insulin cycle)

Cystine

An essential, Sulfur-containing amino acid, produced by digestion or acid hydrolysis of proteins, sometimes found in the urine and kidneys, and readily reduced to two molecules of Cysteine.

- Reported helpful in dermatological conditions.
- Promotes faster recovery of tissue after surgery.
- Part of the Insulin molecule.
- Found high in hair (sulfur bonds).

GABA - (Gamma Amino Butyric Acid)

An amino acid that is one of the principle inhibitory neurotransmitters in the central nervous system.

- Useful in schizophrenia and epilepsy, depression, high blood pressure, high stress levels, manic Behaviour, and acute agitation due to increased regulation of nerve firings and enhancement of Niacinamide binding receptors.
- May induce calmness and tranquility by inhibiting neurotransmitters, which decrease the activity in neurons.
- May be useful in reducing an enlarged prostate due to the suppression of Prolactia released by the pituitary gland.

Glutamic Acid

Glutamate is a salt of Glutamic Acid; in biochemistry, the term is often used interchangeably with Glutamic Acid. Glutamic Acid is a crystalline dibasic nonessential amino acid, widely distributed in proteins, which is thought to be a neurotransmitter, inhibiting neural excitation in the central nervous system; its hydrochloride salt is used as a gastric acidifier. The monosodium salt of L-Glutamic acid (sodium glutamate) is used in treating encephalopathies associated with hepatic diseases, and to enhance the flavour of foods and tobacco.

- Reported useful in Muscular Dystrophy.
- Does not cross blood brain barriers.
- It is an excitatory neurotransmitter in the CNS; (Precursor of GABA.)

Glutamine

The monoamide of Glutamic Acid, an amino acid occurring in proteins; it is an important carrier of urinary ammonia and is broken down in the kidney by the enzyme glutaminase.

- Is useful in treatment of Alcoholism by reducing the desire to drink.

- May be helpful in the improvement of Autism and mental retardation in children.
- Reported improvement in mental functions such as memory and dexterity.
- Aids peptic ulcer healing due to antacid quality.
- Crosses Blood brain barriers.
- Participates in nucleonic acid synthesis.
- Converts to Glutanic acid.

Glutathione

Reduced Glutathione, a tripeptide of Glutamic Acid, Cysteine, and Glycine, which serves as a reducing agent in many biochemical reactions being converted to oxidized Glutathione (GSSG) in which the Cysteine residues of two Glutathione molecules are connected by a disulfide bridge. Reduced Glutathione is important in protecting erythrocytes (red blood cells) from oxidation and hemolysis; deficiency causes sensitivity to oxidant drugs.

- It is a tripeptide containing amino acid that contains: Cysteine, Glycine, and Glutamate.
- Inhibits peroxide formations, reducing free radical damage.
- Detoxifies aromatic hydrocarbons common in air pollution, such as chlorine. Reported to have protective function against radiation therapy.
- Transports other amino acids into the interior of the cell.

Glycine

A nonessential amino acid, occurring as a constituent of proteins and functioning as an inhibitory neurotransmitter in the central nervous system; used as a gastric antacid and dietary supplement, and in the treatment of various myopathies.

- Is the simplest and sweetest of the amino acids. Can be used as a sweetener in herbal beverages.
- Reported useful in degenerative diseases such as MD (Muscular Dystrophy)

- Detoxifies benzoic acid (a common food additive) and aromatic hydrocarbons in the liver.
- Involved in synthesis of nucleic acids and bio-acids. May be useful in conditions characterized by abnormal nerve firing such as epilepsy, inhibition of tripeptides Glutathione and GTF.
- Is one of the amino acids in the tripeptides Glutathione and GTF; low brain concentrations of Glycine have recently been found in ALS. (Amyotrophic Lateral Sclerosis or"Lue Gehrig's Disease.")

Histidine

Is not listed as essential for adults, but is very essential for infants. An amino acid obtainable from many proteins by the action of sulfuric acid and water; it is ESSENTIAL for optimal growth in INFANTS. Its decarboxylation results in formation of histamine.

- Release of histamines from body stores are required for sexual arousal.
- Reported useful in alleviating pain associated with rheumatoid arthritis. (Should be taken with vitamin C; Should be used with caution with Manic Depressives with elevated histamines; Should be used with caution in woman with severe depression or suicidal tendencies due to Premenstrual Syndrome; Is considered a mild neurotransmitter).

Hydroxyproline

An amino acid produced in the digestion of hydrolytic decomposition of proteins, especially collagens.

Isoleucine

An essential amino acid produced by hydrolysis of fibrin and other proteins; ESSESNTIAL for optimal INFANT growth and for nitrogen equilibrium in adults.

- Needed along with other branch chain amino acids for all rebuilding of muscle tissue.

- Role in the release of energy during muscular work. (Metabolizes along the same pathway as fat.)

Leucine

An essential amino acid, ESSENTIAL for optimal growth in INFANTS, and for nitrogen equilibrium in adults.

- Metabolized along the same pathways as fat.
- Precursors of cholesterol.
- Involved in the role of energy release during any work of the muscles.

Lysine

An essential, naturally occurring amino acid, ESSESNTIAL for optimal growth in human INFANTS, and for maintenance of nitrogen equilibrium in adults.

- Reported to inhibit growth and replication of herpes simplex and Epstein Barr viruses (EBV).
- Promotes bone growth in infants.
- Stimulates secretion of gastric juices.
- Found in abundance in muscle tissue, connective tissue and collagen.
- Low in vegetarian diets.

Methionine

An essential, naturally occurring amino acid, which is an essential component of the diet, furnishing both methyl groups and sulfur necessary for normal metabolism.

- Prevents deposits and cohesion of fats in the liver.
- Gives rise to Taurine (an important inhibitory neuro modulator in the brain).
- Involved in synthesis of Choline (must be given with Vitamin B6 (paradoxial-5-phosphate) to inhibit synthesis of homocysteine which promotes plaque deposition in the arteries)

Autistic Patients

"Taurine is deficient in Autistic patients; Methionine has

an altered abnormally normal to high level, with anything distal to the Methionine pathway being low. Since Taurine is so important in the intelligence quotients of most species, this seems to be a significant.

" Alpha A - Leverton, Protein and Amino Acid Nutrition (1959) B - Rose, Journalism of Biological Chemistry 217:977, 1955.

Ornithine

An amino acid obtained from Arginine by the splitting of urea; it is an intermediate in urea biosynthesis.

- May reduce fat and increase muscle mass by promoting fat metabolism and stimulating growth hormone production.
- Helps in detoxification of ammonia in the urea cycle.
- May be useful in autoimmune disease such as arthritis.

Phenylalanine

An essential, naturally occurring amino acid, ESSESNTIAL for optimal growth in INFANTS, and for nitrogen equilibrium in human adults.

- May be useful in appetite control by stimulating CCK (Chaleceptokinin Enzyme) secretion.
- Shown to be useful in management of certain types of depression.
- Increases blood pressure in hypotension.
- Gives rise to Tyrosine
- It is one of the amino acids in the dypeptide sweetener Aspartain.

Proline

A cyclic amino acid occurring in proteins; it is a major constituent of collagen.

- Is an anti-hypotensive agent in lowering high blood pressure.

- Reported helpful in repairing muscle and tendon damage.
- Reported useful in promoting skin flexibility in relation to aging and sun exposure.
- A major amino acid found in collagen (connective tissue) in the presence of Vitamin C.

Serine

A naturally occurring Amino Acid, present in many proteins.

Taurine

A crystallized acid, ethylamine sulfonic acid, from the bile; found also in small quantities in lung and muscle tissue.

- Low levels seen in newborn infants fed low Taurine diets.
- Associated with retinal degenerations.
- The role of Taurine as a nutrient is to protect the cell membranes by attenuating such toxic compounds as oxidants, secondary bioacids and antibiotics.
- Recommended for children on long-term parenteral nutrition.
- Helpful in balancing calcium and potassium flux in heart muscle.
- Helps patients suffering from congenitive heart failure by alleviating their physical signs and symptoms.
- Increases left ventricular performance without any significant changes in atrial pressure.
- Often times considered a neuro modulator.
- Helpful in treating some types of epilepsy.

Autistic Patients:"Taurine is deficient in autistic patients; Methionine has an altered abnormally normal to high level, with anything distal to the Methionine pathway being low. Since Taurine is so important in the intelligence quotients of most species, this seems to be a significant.

Threonine

An essential, naturally occurring amino acid, essential for human metabolism.

- Rises to three times it's normal value at pregnancy.
- Acts as a lipotropic factor.
- Recently found to increase brain Glycine content, greatly reducing ALS symptoms.

Tryptophan

An essential, naturally occurring amino acid, existing in proteins and essential for human metabolism.

- Reported useful in the management of depression and schizophrenia.
- Produces Serotonin, which induces sleep. Has a Serotonengenic effect.
- Precursor of the vitamin Niacin.
- Vasoconstrictor, which appears to aid in blood clotting mechanism, aids in elevating the threshold of pain.

(Should be taken with Vitamin B6 (paradoxial-5-phosphate) and in the presence of carbohydrates such as fruit or vegetable juice to maximize uptake in the brain.)

Tyrosine

An essential, naturally occurring amino acid present in most proteins. It is a product of Phenylalanine metabolism and a precursor of thyroid hormones, catecholamines, and melanin.

- Along with Phenylalanine, it is a useful anti depressant due to increased production of Tacolamine. (Recommended to be taken with Vitamin B6 (paradoxial-5-phosphate).)
- Reported to stabilize blood pressure by lowering energy in some cases and elevating it in others.
- Involved in tissue pigmentation
- Is important in the formation of thyroid hormone

(Should not be used when MAO's are prescribed or when cancer melanoma is present)

Valine

An essential, naturrally occurring amino acid, essential for human metabolism.

- Needed for all muscle building.
- Required in the precursors of cholesterol.

Index

G

H

I

K

L

M

N

O

P

R

S

T

V